MORE PRAISE FOR V

"*Virology* is a brilliant book, both playful and serious, showing us all how viruses live with us, as we live with them. Drawing on queer theory, Joseph Osmundson offers a way of understanding care in the midst of anguish and anxiety as well as desire and hope. The viral world is the ordinary world of life and death, of caring for one another in our vulnerability and persistence. This book explains the science of virology for our times, offering a compassionate education for all of us disoriented in pandemic times. This book is queer pedagogy at its best: non-patronizing, thoroughly smart, and full of urgent and caring knowledge that beckons us to get closer again with caution and passion." —Judith Butler

"The essays in *Virology* are beyond impressive. This is precision work, cutting and thoughtful, done with the deft hand of a wildly skilled writer. Osmundson has given us something precious with this important collection. It is a tribute to humanity. It is an ode to life." —Kristen Arnett, author of *With Teeth*

"Joseph Osmundson's *Virology* is an incisive look at our relationship to earth's most plentiful life form—how we live with viruses and how viruses live in and through us. But more than this, it is a compelling examination of the tension between avoidance and exposure, safety and risk, preservation of the self and openness to evolution and change. This book is a potent medicine for our times."
—Lacy M. Johnson, author of *The Reckonings*

"Joseph Osmundson's *Virology* made me gay for viruses. Seriously. *Virology* is a tour de force that uses queer theory to teach us about the science of viruses. Along the way, we are forced to reckon with the reality that far from being villainous little creatures, viruses are actually fascinating almost-life forms. *Virology* brilliantly revises the frameworks we use to talk about life in a world filled with viruses and reminds us that our relationship with science and scientific phenomena is always social."

—Chanda Prescod-Weinstein, author of *The Disordered Cosmos*

"I have absolutely no idea how Osmundson made a book this timely, this timeless, this packed with contents and styles we aren't supposed to experience in one text. *Virology* is devastating in its soulful brilliance. Rigor just became cool as fuck and pleasurable again."

—Kiese Laymon, author of *Heavy*

"Inquisitive, bold, and lyrical, *Virology* offers a captivating and very queer look at our present moment through the lens of someone who knows more than most of us about the science behind our shared catastrophe." —Melissa Febos, author of *Girlhood*

"To read Joseph Osmundson's mind at work is such a pleasure. The tendrils of *Virology* go deep: to the pandemic, queerness, memes, futurity, and what it means to hold both love and despair, to live awake to both the world's beauty and its harm. This is a profoundly necessary, urgently of-the-moment collection, one I'll keep thinking about for a long time to come."

—Alex Marzano-Lesnevich, author of *The Fact of a Body*

"Luckily, we have Joseph Osmundson—an actual virologist—who writes with elegance and insight about the intersection of the real and the metaphorical, moving through topics like the legacy of HIV/AIDS, the long-term impact of COVID variants, and the effects of a prolonged pandemic on our systems of power. This is very much a book of our times." —*Literary Hub*

Virology

ALSO BY JOSEPH OSMUNDSON

Capsid: A Love Story

Inside/Out

VIROLOGY

Essays for the Living, the Dead, and the Small Things in Between

JOSEPH OSMUNDSON

W. W. NORTON & COMPANY
Independent Publishers Since 1923

Printed in the United States of America
First Edition

Manufacturing by Lake Book Manufacturing
Book design by Beth Steidle
Production manager: Beth Steidle

ISBN: 978-0-393-88136-3

W. W. Norton & Company, Inc.
500 Fifth Avenue, New York, N.Y. 10110
www.wwnorton.com

W. W. Norton & Company Ltd.
15 Carlisle Street, London W1D 3BS

1 2 3 4 5 6 7 8 9 0

For Randall, my first teacher.

SARS-CoV-2, like HIV, is a virus that is transmitted between individuals in the course of human relationships that take place in social and environmental context.

—JUDY AUERBACH

bc there's my body
and then there's your body,
and I don't think anybody's
coming over tonight.

—TOMMY PICO

Contents

Virology

1

On Risk

What Is There to Fear?

It's March 16, 2020, about a week before New York State will go on pause. I wait outside of Trader Joe's on Manhattan's Lower East Side, all post-industrial buildings, all gray and concrete, and the sky gray too, low and spitting rain. Fuck. I forgot my umbrella, too late to go back for it. I figure that on a Monday afternoon traffic to the store will be low.

I stand in the Trader Joe's line outside under the spitting sky wondering what the store will be like inside. Crowded? Will there be lines for checkout too? There's no tape or chalk on the ground marking 6 feet. But I know enough to stand 6 feet behind the person in front of me and to stare at the person behind me with an intensity that keeps them 6 feet behind me too. I'm treating everyone like they might be infected, because they might be, and I'm treating myself the same way too. I'm a scientist. This is what I know.

SARS-CoV-2 is similar to other coronaviruses in how long it can survive on surfaces. What a virus is made of dictates how it will act. Coronaviruses are enveloped, wrapped in membrane just like our cells. Their genetic information is made of RNA, the less stable cousin to our cells' DNA. And yet, these little sacks of membrane and RNA and a few proteins, too, can still remain infectious on certain hard surfaces for a few hours.

Riding down the escalator into Trader Joe's, I'm shocked at how empty it is. Keeping the line outside, where wind can whisk viruses away, that's smart. Suddenly I see a virus on every surface. That pile of avocados. That freezer section with only three pizzas left. The red onions I'm stocking up on. From every surface, that virus gets on my hands. But I know enough to know: It can't infect me from there. I do not touch my face. I use the back of my hand to move my glasses up my sweaty nose. I think about José, I wish he was still here.

José Esteban Muñoz was a Cuban American scholar of queer studies before he died in 2013 at only 46. He wrote two books before he died, *Cruising Utopia* and *Disidentifications*. I often find myself mourning the voices that I wish we still had to write us through a present crisis, even in line at the grocery store. Building on work by Leo Bersani, Muñoz wrote that because queer people may choose not to have biological children, queers are "within the dominant culture, a people without a future."

I've argued for some years that climate change may make us all queer, a worldwide people without a future. For now, though, this is inverted: we are all without a present, living only for a future when things won't be like this. We're locked inside with our lives on hold. I can barely stand the stress of coming here, to Trader Joe's. I stare out the window all day as I work, my only access to the outside world beyond this weekly trip to the grocery store.

This, for me, seems akin to queer childhood, growing up in tiny towns just living, just surviving, on the hope of growing old enough to leave.

For queer people, the life offered to us is never enough. And queerness isn't (just) who we sleep with; as feminist scholar bell hooks explains, "queer as not about who you're having sex with—that can be a dimension of it—but queer as being about the self that is at odds with everything around it and has to invent, and create, and find a place to speak, and to thrive, and to live."

Queer childhood is waiting for the possibility to be—to *make*—one's full self. Quarantine is putting the full possibility of social relations—one way to *make* oneself *with* others—on hold out of respect for the desire of living beings to keep on living.

In his book *Image Control*, Patrick Nathan argues that climate change is making us futureless, living at the end of the world as we've known it. A pandemic makes us live without a present, putting life on hold for the future when things will be better. No wonder 2020 presented such a challenge to us all, living without a present and understanding, too, the limits of our future.

When all we wanted to know was when will we be able to—once again—go outside without calculating probabilities of disease, death?

The basket of groceries digs into my hand. It's the liquids, I know: seltzer water, iced coffee concentrate, some beers, because nothing matters anymore at the end of the world. There's no line to check out. I stand 6 feet, as best I can, from the nice woman running the register, number 17. She discovers that I teach at NYU, that I studied microbiology, and now she wants to talk.

"What the fuck is going to happen?" she asks.

"I don't know, but it's going to be like this for a while," I answer. "How are you holding up?" I ask.

"It's not that bad," she says, and I look at her sideways. "OK, fine. It is. But what else can I do?" I nod, guilty for having put my body so close to hers. I pack my groceries into my massive camping backpack. I walk up the escalator, it's empty, no one within 6 feet of me in any direction.

Staying inside is painful, it's mentally exhausting, it . . . hurts. ("The pleasure and pain of queerness are not a strict binary.") And yet, if we invert the mix of boredom and terror that is quarantine, we can understand it as a kind and generous act, one that gives us pride and, through pride, joy. By staying inside, I'm caring for myself, yes, but also for this woman, an essential worker, worried about her own risk too.

"The queerness of queer futurity . . . is a relational and collective modality of endurance and support." Being queer is a legacy and a history of care even in the face of systemic oppression, violence, murder, viral disease. Queer people have been training for this moment—to sacrifice, in the face of a virus, to care for one another, and yet to never lose sight of pleasure, even when both the present and the future seem impossible. It is too much to ask of us. What choice do we have but to do it?

....

The first time I went down on a guy, my mind was doing math. One in 12,000 was the risk to contract HIV, I knew. "Oh God, this is a mouthful," I thought. I was drunk, but my work as a scientist involved doing math even while inebriated. I didn't know what exactly it tasted like, but I sure knew I liked it. It tasted more like a body than anything I'd had up until then. I didn't know why, but I grew hard at the tasting. I knew, too, that the virus, that virus, was poorly transmissible in any single act. Even bottoming (receptive anal sex) without a condom only had a 2 percent chance of seroconverting—moving from the binary category of HIV-negative to the seropositive category of HIV-positive.

I was—I am—a professional biologist. I did my postdoc, in part, in biostatistics. Long before I touched another man, I'd gone to the CDC website and looked at a table of acts and percentages, calculated years ago.

My mind plots data without thinking about it, likelihoods over time. When it comes to viruses, I see curves of two types, one with time on the *x*-axis and a line on the *y* approaching 100 percent over time: the likelihood of being infected with a common human virus, like cytomegalovirus, or CMV, a part of human experience. Between zero and two years, we *will* meet it. The other curve, the rare deadly virus, the one we do anything and everything to avoid—there my risk behaviors on the *x*-axis and a line on the *y* approaching zero, no risk.

I know I will never arrive.

This is what drives my nightmares, my catastrophizing. Probabilities and statistics have a wicked sense of humor. A 2 percent chance seems unlikely until you end up on one side of it, the rarer side, asking how exactly is it that you arrived at this particular, unlikely, destination.

"But I only did it once . . ." I imagined myself saying to keep myself from doing it at all. Every time I got naked in proximity to another man, my mind was a calculator. I'd spent my life afraid of HIV and so never able to rid myself of worry or of stigma for those who had it. I'll never not be a queer person born in 1983, born into what viruses meant then. We know now that the only risk-free sex is with an HIV-positive person who's undetectable, keeping their viral load at near zero with drugs. I've had sex with HIV-positive men, and my mind arrived only at one round number, the one I love most: 0, wide open like a mouth.

I've been obsessed with viruses ever since I read *The Hot Zone* in middle school. I wanted to be a virologist at the WHO when I grew up. This is why I'd studied French, and not Spanish, in high school. French introduced me to a new type of literature, one that didn't just tell stories but questioned the relationship between stories and reality, and what was reality anyway? I never did go work in Geneva. I used my French in my work as a scientist only once, on a year abroad before starting graduate school, studying the biophysics of the prion protein and mad cow disease in a hospital lab in the French Alps.

My mom reminded me that I'd wanted to be a virologist my whole life. A virologist or fireman, is what I used to say. Guess which one I dressed up as for Halloween? Too poor to buy costumes, my parents modified my dad's firefighting outfit that he used in the summer to fight forest fires. The outfit smelled of him, and of ash. I'd grow up to wear what they couldn't afford then to buy: a lab coat, goggles, and gloves.

"You're doing what you always wanted to do," she said at my PhD defense, but I'd forgotten that particular desire over the course of my life. I'd moved on to wanting to be a lawyer, a doctor, a professional soccer referee. I'd managed to somehow come back to viruses.

Thanks to the coronavirus, Americans are inundated with numbers in a way that's rare outside of political punditry. The mortality rate is 3.4 percent, or 5 percent, or 1 percent, or, according to then president Trump's statements from early March 2020, 0.1 percent, just like the flu. We thought the R_0 was between two and three, but now we know it's actually between five and six, meaning that each infected person will—on average—infect five to six other people. And this was before the delta variant. The doubling time for the number of infected individuals is around five or six days, which also, confusingly, happens to be COVID-19's average incubation period. The virus can live on surfaces for 12 hours, or 2 days, or 17 days if on a cruise ship, apparently.

We all just want to know: Is it safe to go outside? Can I . . . hook up? We're all so stuck in the minutiae of risk, we're all so inundated by data that even the experts can't yet make sense of, that we have no idea how in the world to act.

....

What is the risk to me? I'm almost 40 with no preexisting conditions. Reports state this virus can be nasty, if not usually deadly, for people like me. But young people in America are dying of COVID-19. And people of all ages can become severely ill and require hospitalization to live. Some people stay ill with COVID-19 for months, including young people. And if our hospitals aren't functional because there are simply too many cases, the mortality rate for this virus can climb, as it did in areas of Italy.

Remember, statistics are darkly humorous. Risk spreads itself out across the world, to everyone.

Care does not require anxiety or punishing ourselves for what we cannot avoid. Care does not require perfection, as that's something no human body can attain in this moment.

I feel guilty for having gone to the store. I put myself at risk, and while that's bad, there's something worse. I could be carrying the virus, but not feel sick, and depositing it wherever I go.

I feel guilty, but I have a mouth to feed. My own. I'm still human. My O-shaped mouth. As a human, I'm a walking SARS-CoV-2 incubator. If the virus wants anything, it wants me.

We're human, which necessitates a world outside of our apartments. Anything out there, in that world, brings the possibility of illness. I washed my hands. I washed my veggies. I put rubbing alcohol on the bridge of my glasses, where the back of my hand always pushed them up, out there, in the world.

The risk is not, and can never be, zero. Could I order groceries, have them delivered? Sure, of course I could. And that would push some of my risk onto another person, a shopper in the store, a delivery person on a bike. That person is more likely to be a person of color and more economically at risk than I am. My risk may be lowered, but the overall risk remains, and will always remain, as long as my body has needs, which is to say as long as I live.

We have to reframe the very notion of risk, of fear. It can never be zero. The more we *all* minimize risk, the less there is to fear.

....

"Concrete utopias are the realm of educated hope," Muñoz tells us; they are "a backward glance that enacts a future vision."

How can we speak of utopia in the time of plague? Utopia, in Muñoz's imagination, doesn't mean a world without harm or hurt or risk; that world wouldn't allow the possibility of human life. No. Utopia is a world where risk is minimized for as many people as possible, where we care for every life, even queer ones.

Utopia is not former president Trump's manipulative optimism, but rather a knowledge that, whatever the future, we will care for one another through it all. Manipulative optimism is the stuff of abusers. They convince you to accept a horrific present with the promise of a perfect future (COVID will be gone by Easter!) that we all know will never come. Educated hope means not turning away from evidence, as Trump consistently did, but using evidence to guide us toward the best path forward—out of the past, through this present, and on to a better, more just world.

When Jacques Derrida was asked, near his death, what he meant when he called his work a "writing of survival," he answered, "If I had invented my writing, I would have done so as a perpetual revolution." Muñoz's notion of queerness is just this, a perpetual striving toward something that doesn't yet exist in the world, but that we can feel, in our guts and bodies, is necessary. Queer people in a straight world are always pinched, unable to be fully or freely ourselves.

Muñoz wrote, "The here and now is simply not enough." Especially not now. I can't even go to the store when I want, much less a bar with friends, a club in which to dance. "Queerness should and could be about a desire for another way of being in both the world and in time, a desire that resists mandates to accept that which is not enough."

We won't accept this world as it is now, or the one that was before. We embrace, but do not accept, our chosen self-confinement. We do this because we know it will save lives, and lives are worth saving, even if it hurts.

....

My risk and yours, our personal risks, will always be more than zero. There's a solace in this. If we're all doing the best we can, then it's no one's personal fault if they happen to fall ill. We know that people are sick now, and more will get sick. This is a virus, and that's what viruses can do. This virus will be with us for weeks and months, maybe years, maybe, in some ways, forever—what virologists call endemic.

So be queer. Build nonnuclear families of mutual support and care. Be together in person if you live together or near. Drop off food. Skype. Sext. Send food. Zoom. Send nudes.

God, grant me the serenity.

To live queerly. To embrace the small pleasures of the here and now, this moment of forced isolation. This impossible present is training for our impossible future.

I've stopped running calculations. Instead, I run outside on the East River, measuring myself 6 feet from other bodies. Most days, I don't even go outside, I run up and down the stairs in my apartment building, six flights to the skylight on the last. On my first trip up, I wondered why the light got brighter with each floor I ascended. I didn't even notice the sun until the crisis made me run stairs, pull myself up toward the heaven, protected still by glass.

Returning from Trader Joe's, I wash my vegetables in water

only after I've soaped my hands for three singings of "Happy Birthday." I'm not doing math. I just know, in my gut, I killed 99 percent of the things on me, and that's as good as I can do. I can't spend time mourning the fact that Muñoz isn't here any longer to guide my thinking. It's high past time that we step up and continue his work together. I'll keep doing as good as I can do until the end of this crisis, and then still after that, until there are no more crises left to live through.

2

On Replication

An Introduction

MYOVIRIDAE (BACTERIOPHAGE)

Examples: Phage T4, Peduovirus (P2), Twort, phage G1

Look with me.

You're swimming, out past where your toes can touch, salt water between your feet and the sand. The day is sunny, the air wet and warm, and the cool water does you good. Imagine the world, a new world. You can see the very small. Your eye, a microscope. You can see inside a cell. Your eye, an electron microscope. You can see even smaller than that. And what do you see? A virus, a virus, a virus. There are 250 million viruses in every 0.001 liters of ocean water, and so 7,393,387,354, more than 7 billion viruses, in 1 single fluid ounce, a mouthful, so much salt water, spit it out. Multiply that by the volume of ocean on the planet, my God. A swell, coming your way, dive under, hold your breath but open your eyes to the blue-green of the underwater realm of our one earth. How many viruses there must be in that wave, pulled by the moon, coming your way.

But worry not, dear, these viruses in the ocean only infect bacteria; they can't touch you. Lie out on the beach, let the sun dry the salt water and its viruses into a white powder on your skin. Take a selfie. Zoom in. Zoom in. Zoom in. Get to a point where you can see your skin, really see, the tissue, dead cells sloughing off at the surface, peel them back, dig deeper, it'll only hurt a little. Can you see your own cells? Bags of liquid and protein, bags of DNA and its cousin molecule RNA.

The closer we look at a virus, the more meaning we can make from it. That virus. There. HPV, CMV, HIV, MMLV. Will it be with me a day, a week, forever, never? Will it kill us or make us itch or make us sneeze or make us sick? Look to its molecules, gaze at its surface, and we can together start to answer these questions.

Our world is built of things we can't see. The bodies we inhabit are made of the invisible. I see skin, I see hair, I can look into the eyes of a human and feel something move me, knowing that the eyes, the hair, the skin are nothing but ordered, repeating units of protein, of fat. Thought itself is salt moving fast across fatty membranes in the cells in our brain and body.

"Most of what matters in our lives takes place in our absence," Salman Rushdie writes in *Midnight's Children*. We take place in our absence, too. All decisions we make begin in the cells—which we cannot see—of our brain—which we cannot see. Our thoughts and feelings originate from our body's invisibility.

For hundreds of years, scientists have convinced us that things too small to see not only exist but make up all life on earth. It's a story I wouldn't believe but one I know to be true. Have faith; we've invented ways to see units of protein, units of fat, that build our hair and eyes and skin. Stare long enough at the night sky and it is apparent that the earth moves around the sun.

Viruses were identified as an invisibility, something that could kill, which we could see, but a microscope strong enough to see the virus itself was yet to be built by human hands. Yet we know their material consequences, we know them to be real, so real they can kill. Childhood's chickenpox left a scar on my chest, left side, outside my heart, where I scratched a lesion until it bled, and then pulled off the scab and scratched it again, for a week on end.

Viruses, invisible, are the most abundant life-form on earth.

A virus to a cell is nothing without an entry point. A virus is not alive, because one of the central tenets of life—to a biologist—is the ability to self-replicate. To copy oneself. This is true of my cells, and so they are alive. One cell can copy itself into two. Alive.

The virus will need some cell, something besides itself, to reproduce, to replicate. It's not living, it's not dead. Bacteriophage, or phage, are viruses that infect only bacterial cells. To copy themselves, they need bacteria just like human viruses need me. The deepest desire, something like love. We complete it.

I'm a doctor because of these phage. In my PhD, I studied phage G1 and phage Twort, two phage that infect and kill staph bacteria. These viruses couldn't touch me. I could grow them up in the lab without even wearing gloves, goggles, a lab coat. I could have swum next to them in the ocean; if I swallowed the water, the salt would be more dangerous than the phage. I could have picked them up, these viruses, in a handful of soil. I wanted to learn from the viruses how this bacteria could be killed. I owe them so much, these G1 phage, bacteriophage to put it formally, family *Myoviridae.*

Bacteriophage from the family *Myoviridae* are the most common viruses on and in us, so much more common than the cold or COVID. The most common viruses on earth. The most common life-form on our planet. On our skin, in our gut, in our nose and mouth, in our lungs and piss. Between 100 million and 1 billion virus particles (mostly bacteriophage) are present in every gram we shit, and the average human shit is 225 grams or so. Our gut is a viral fermenter.

We scientists classify viruses by the material they use to encode their genes (with DNA and RNA being the two options) and whether or not they wrap themselves in a fatty, lipid outer layer, like our cells do.

Our genome—the collection of genes we have that, in many ways, make us—is 3.2 billion letters long. Our genome is made of double-stranded DNA, so iconic it's an emoji. Everything alive writes itself down in DNA. Viruses, not living and not dead, have so many more possibilities.

DNA has four letters that encode information: A, G, C, T. These represent different bases, different molecules, in addition to a sugar and phosphate backbone that repeats. Our DNA exists in 23 unique chromosomes, 23 single extended molecules, each a long sentence of letters. These letters can code for proteins, made of amino acids put in a particular order. Proteins in our cells fold up into unique shapes that perform certain functions: they hold the cell up; they are enzymes that process energy from the sugar in our food into mitochondria, ATP from mitochondria that our cells can use to move, to lift, to think.

This order of information, from DNA, where it's stored, to RNA, where it's accessed, to proteins, which enact a function, is so fundamental to life that biologists named it the central dogma. That living things store their information in DNA is how life works, and there are—in life—no exceptions.

But viruses aren't truly alive. They cannot replicate on their own. They need something. Someone. They need *us.*

Viruses have genomes too, little recipes for themselves. Everything they need to copy themselves, with the help of our cells, is right there. Mostly, these genomes are small, tiny compared with ours. If the genome of HIV, at around 10,000 bases, for example, is the area of the period at the end of this sentence, our genome would be around the area of the cover of a microbiology textbook, a foot squared or so.

Myoviridae phage are unenveloped, naked, double-stranded DNA viruses. Viruses, both those with layers of membranes around them and the naked ones, all contain capsids, or protein shells. So enveloped viruses have two layers, two shells, one made of fat outside and one made of protein inside.

Phage in the family *Myoviridae* have genes made of DNA. They contain 100 to 400 genes. Big for a virus. The bacteria they infect have something like 5,000 genes, and yet these little bugs can kill that bigger and more complicated thing. Us? We have around about 20,000 genes or so.

It doesn't take many genes to replicate, after all, if that's all you're trying to do. Most viruses are compact, keeping only the genes they absolutely need. As biologists, we can almost never say anything is always anything; as soon as we say "anything" without qualification, an exception will pop up and prove us a fool. I've always found this humility a valuable life lesson.

These viruses, phage, don't have fatty membranes like our own cells, and are shaped like moon landers, made of protein alone. Once they've landed on bacterial cells, phage squat and inject their DNA inside. Once that barrier has been crossed, their DNA now in the bacterium, they use all the machines the bacteria has to copy those genes and make proteins. They steal the machinery from the cell, copy themselves by using the same processes as the host bacterial cell, make thousands of copies of themselves, and then—oops—excuse themselves, burst the cell, and leave, goodbye and good night.

The way a virus encodes its genes limits its options for replication. Myoviruses have double-stranded DNA that looks just like the DNA of the bacteria they infect. Hello, dear moon lander, thank you. You help make the world, and I'm grateful to you. Hello, dear protein shell, I love you. I spent six years, more, thinking about you every day, and what is that but love? Goodbye, tiny virus, I spit you out with so much salt water. You reminded me, when I needed it most, that things so small they're invisible can still be beautiful, like a thought or a feeling (nothing but salt moving in my mind), like you, dear virus, dear phage, in me, even now.

ADENOVIRIDAE

Examples: Human adenovirus group D type 13,
human adenovirus group D type 8 (HAdV-D type 8),
HAdV-G type 52, Sturgeon ichtadenovirus A

Replication, derived from the early English (fifteenth-century) word for "repeat," itself from the Latin *replicatus*, the past participle for "fold back, fold over, bend back." *Re-* is a common prefix indicating "again," and *plicare* denotes a folding.

Growing up, my mom baked bread every Sunday for our one ritually enforced family dinner. No soccer practices or band lessons or friend sleepovers were allowed to stand in the way of family dinner. And with the day ostensibly off work, although there always seemed to be work in the garage and garden and backyard, bread could be mixed in the morning and left to rise before being kneaded in the afternoon, left again to rise, and then baked in plenty of time for a 5:45 p.m. meal.

We let the dough rise again to develop its consistency, those nice air pockets. That air comes from the budding yeast, which grows quickly, doubling the number of cells in the bread culture every 90 minutes or so, converting the sugars in the flour into energy in their cells, producing carbon dioxide gas as a by-product.

Each cell division makes two yeast cells from one. Each division requires copying the cell's DNA such that each daughter cell gets its own copy.

We call this replication.

Yeast take 90 minutes to do all the work of making two cells from one. Human cells take 8 hours to copy their DNA alone. Viruses, in 8 hours, can make hundreds of thousands of offspring, their replication taking only a few minutes to complete.

Viruses don't follow the rule of cells. Cells divide one and two and four and eight and sixteen, each cell turning into two. There's no way to go faster than that. No matter how rushed we are, we have to let the bread proof for an hour.

Viruses can make thousands of offspring in a single round of replication. Viruses need us to replicate, but with us, they can do things we still find impossible.

Our cells: one, two, four, eight. A virus: one, ten thousand, one hundred million, and like that, the virus has done its work, has replicated and is ready, now, to move on, to infect another cell or another body.

Adenoviruses, like us, make their genes from double-stranded DNA, that famous double helix, so permanent. Like phage, they have no envelope of membrane around themselves. Unlike phage, they can make us their hosts.

Because adenoviruses—named for the adenoid tissue they were first discovered in—are made of the same material we are, once they get inside us, and once they get inside our nucleus, they can start replicating themselves, just the way a phage might do in a bacterium. Adenoviruses have proteins on their surface that can bind to a protein on our cells, their doorway inside. What are you made of? What do you have on your surface? Answers to these questions define so much about a virus.

The first step, for a virus, is gaining entry to a cell, the factory it'll use to xerox—to photocopy—itself. Any virus needs to bind to something on the outside of our cell to attach and then somehow break past the cell's membrane to come on in, to get out of the cold outside world. Different adenoviruses use different proteins we have on different cells; this is one way that different viruses can cause different diseases. Viruses that attach to lung cells cause diseases like pneumonia; viruses that bind to cells around the eye cause conjunctivitis, pink eye. Adenoviruses can replicate in the lung or the eye; they can cause disease, yes, but most of the time, in those folks with an active and working immune system, the virus itself goes entirely unnoticed by the human precisely because it's noticed by the cells in their immune system. No pink eye, no cough or shortness of breath. Viral replication occurring, yes, but not so much that it makes us ill.

An adenovirus may be replicating inside me even now, common enough as it is. In one study of children, up to 80 percent of kids had an active adenovirus in them. I've had pink eye, and I don't want it again, I'm sneezing now with my fall allergies, my own cells now replicating as they always do, my yeast cells growing away in my lab, replicating as I told them to.

HERPESVIRIDAE

Examples: Cytomegalovirus, Epstein–Barr virus, herpes simplex virus 1 (HSV-1), HSV-2

Herpesviruses infect us all. There are herpesviruses that infect birds and fish, reptiles and mollusks. To be human is to have herpesviruses inside our cells replicating alongside us, to be a human/virus hybrid almost from the year we're born.

Philosophers, too, are concerned with this question of replication and have been as long as philosophy has existed. Plato defined different kinds of replication, one being true to the original object, and another being a representation, obvious in its difference, but that points to the original object still. French postmodern theorists extended this thinking, defining a simulacrum, a copy of a copy where the original has been lost or no longer exists. The representation, then, is all there is of reality. Or: it distorts reality entirely.

Biology, at the cellular level, complicates these already complicated notions. Most of us think about biological replication at the level we can see with our naked eye: I can make a child, and once that child arrives, it isn't exactly a copy of me, and I'm still here.

But at the level of our cells, one cell divides to make two. Which is the copy? Which is the original? Both daughter cells—as we name and gender them in biology—are copies. Their DNA is replicated by a mechanism where each cell has one strand of the new and one strand of the old, a parent/child hybrid at the molecular level.

Nothing remains untouched. I am 43 trillion cells on a good day. Everything is a copy of the first cell we once were, the one that was a merging of a half-cell from one parent (a sperm) and a half-cell from another (an egg). Every cell of mine is a faithful copy of that first cell. And that first cell, where is it now? Gone, but a piece of it is everywhere so long as I live.

And viruses must unpack themselves entirely to make more copies. A virus binds to a cell and comes apart; none of its offspring will contain the molecules of the parent. They're pure copies, xeroxes made in the host cell by the recipe the parent virus died to pass on, that virus a xerox too. A virus, a perfect postmodern object, an invisible copy of a copy with perfectly ordered capsid edges, an intensely geometrical shape, and more metaphorical meaning than anything of its size should be granted.

Culture is self-replicating, but cannot replicate without us, its hosts. Queer theorists have written for decades on the difference between performance and authenticity, indeed arguing that very little, if anything, can be "authentic." What appears authentic is only what is most rigidly reinforced by norms and mores. As Judith Butler and/or RuPaul remind us, "We're born naked; everything else is drag." José Esteban Muñoz calls this the "fiction of identity," not because fiction and identity have no power. Rather, both are made up by authors as they went along writing, both are acts that combine the agency of an individual with the confines and possibilities and conventions of the world in which that individual is living.

The linear notion of life, from our own birth through our schooling—go to college, get a good job, get married, have kids, put them through college, spend time with grandchildren, you swear they have your eyes, and then die before you become a burden—is defined by Muñoz in *Cruising Utopia* as "Straight Time." Copies and parodies and unfaithful representations "help us imagine new possibilities that interrupt straight time."

Even if I were straight, God forbid, my cells would be queer at least. Failed attempts at perfect copies. Both parent and offspring at the same time. Queerness includes anything that resists simple—nuclear—explanations of family and reproduction, that fucks with generations (parents versus kids) and procreation (mom and dad). My cells, myself, do that every day.

When my cells divide, they give themselves to the process, no copy, no mother, only daughters. Ever new, until I die.

When I think of my cells, all offspring from the moment I was one cell big, I cannot imagine a world less governed by straight time. Few things are linear on the scale of a cell; biology is a science driven by an average of randomly fluctuating events. Chaos drives life.

And that's before we even consider the unliving things that viruses are. Nothing of the original will survive in its offspring. They fast-forward straight time; they grow not linearly, but explode exponentially, replicating usually noticed only by our immune systems, sometimes by inducing a runny nose, a fever, and rarely, something more dangerous to the queer, fragile thing that is human life.

Each virus, when it meets and infects a cell, is infecting the last cell it will ever see. It's the copies of that virus that will have the possibility of a nonlife after.

Herpesviruses are double-stranded DNA viruses like the others we've mentioned so far. Worry not, RNA viruses are to come. Worry not: herpesviruses are unlike the others, too. They're enveloped, for one thing, and so wrap themselves in a bag of fatty membrane they've stolen from the previous cell they've been inside.

What's more, herpesviruses don't come and go. They invite themselves into our cells to stay, for good. Their DNA isn't the temporary DNA of adenoviruses that comes, replicates itself, and leaves. Herpesviruses have DNA like ours: a molecule written for life. Once the DNA gets into our cell, by again binding to a protein on that cell's outer surface, and into our nucleus, it makes a circle of itself, and that circle is maintained. When our cells divide, the circle divides too, and each daughter cell gets a copy. The herpes simplex virus 1 you got from a drinking fountain when you were in kindergarten is in you still, a cold sore waiting to come out whenever your immune system gets beaten down by fatigue, not enough sleep, and too much wine, and cold weather rubbing the skin of your lips raw and exposed.

Adenovirus is acute, a virus that comes and goes; herpesvirus is chronic, a virus that comes and stays forever. If we're all forever infected, are we always ill?

A circular DNA, a viral DNA, an episome, replicating itself, capable of making more virus and killing the cell it inhabits, but just as able to sit and wait and just get passively passed on from one cell to its offspring and the offspring of that cell after that, too. My friend, you're a part of me now, my herpes, my cytomegalovirus, you've been in me from before I could speak aloud my own name. Hating you would mean hating myself, a project I too long invested in.

After the last original creature is gone, a virus will remain, content to be itself, nothing but a copy, waiting still, for its final host.

VIRGAVIRIDAE

Example: Tobacco mosaic virus

Viruses were discovered as an invisibility: a deadly force we couldn't explain. Born from failure, how queer. By the middle of the nineteenth century, research by Louis Pasteur and John Snow gave us the germ theory: the notion that many diseases are caused not by imbalanced humors, but by specific pathogens. One could even see the germs, bacteria, if one looked closely enough under a microscope. One could transmit a disease by transmitting nothing but these germs; if one heated the germs to kill them, the disease would no longer transmit. Infections came from germs; but where did smallpox come from exactly? The germ was nowhere to be seen.

This is the frustrating truth about viruses: they're all so different from one another. HIV from Ebola from herpes from hep and phage and rhino and rota, nora, bunya. What you learn about one virus won't apply to another. They all have different ways to get into cells and, once inside, different ways to replicate. Not knowing which virus may kill us next, it's impossible to know which one to study if you want to save lives. All we can do is study them all, and all we can hope is to learn about life, and if we manage to save lives in the process, we can consider ourselves lucky, we can be grateful that we had the chance to be good.

Almost all viruses are smaller than bacteria, and yet they, too, can cause disease. Unlike bacteria, when they do cause disease, there's nothing to see, at least under a microscope.

It all started with sick plants, not humans. Tobacco plants grown in Europe, not America. Sick plants, touched with spots of gray and black and light green; a disease that didn't kill one plant, but spread throughout an entire crop, ruining it. Research is driven by human interests, now as it's always been. When we are sick, or when we're losing money, we try to understand why so we can fix it.

Two researchers went looking for why: Dmitri Ivanovsky in Crimea and Martinus Beijerinck in the Netherlands. Both started with bacteria, the logical culprit. They touched an infectious plant to another, and lo, an infection. Next they used liquid from one plant and found that it, too, would infect. Finally, they passed that liquid through filters with tiny pores, pores that would trap any bacteria. Still, the liquid infected fresh leaves. The infection only grew on fresh plant tissue, but you couldn't filter it away. Beijerinck named this fluid a *contagium vivum fluidum*, a contagious living fluid, the *vivum* shortened and anglicized to *virus*. The one thing he got wrong, that the fluid contained something living, turned into the name that stuck, that sticks in my mouth even now.

Viruses remained bacteria's unexplained invisible twin for a hundred years. Were they a family of bacteria, just smaller? Maybe a toxin the bacteria made? No one could know because no one could see.

In 1939, with the advent of electron microscopes, there it was, this virus, this nothingness that could kill, this invisibility that was now visible. Based on the experiments, we knew something was there. With new, stronger microscopes, we finally saw: long, perfectly round rods made of viral protein, viral RNA tucked neatly inside, its structure so stable it can be in the sun for hours, so stable it can overwinter on a seed, so stable it can be carried from plant to plant to plant by any insect friend.

Tobacco mosaic virus, family *Virgaviridae*, is the first RNA virus we will consider. RNA, which I've called DNA's cousin, looks almost the same as DNA. In our cells, DNA is that famous double-stranded helix, two opposite strands wrapped around each other, stable. DNA fossilized in a Neanderthal toe some 50,000 years ago was intact enough to be fully sequenced. Writing in DNA, across the scale of a life, is writing for good.

RNA comes and goes; it lives in our cells, but only for a time. Viruses are the only things on earth that still use this molecule for all their writing, that rely on it for their perpetuation, for their replication. Many viruses will die as their RNA degrades. What difference. Little matter. As long as a few survive to replicate again.

Contagium vivum fluidum, an infectious living liquid that isn't alive. Viruses, the most abundant things on earth. It wasn't until the 1940s, bombarded with electrons, that they finally showed themselves to us. Measles, polio, smallpox, phage, and the curlicue capsid of the tobacco mosaic virus. Viruses were more deadly when we didn't understand them. One can only manage what one can easily see.

RETROVIRIDAE

Examples: Moloney murine leukemia virus (MMLV),
ovine lentivirus, feline immunodeficiency virus (FIV),
simian immunodeficiency virus (SIV),
human immunodeficiency virus (HIV)

RNA is often single-stranded and doesn't have the helical structure of DNA, and it's less stable; over time it breaks down. Writing in RNA is writing in a temporary notebook. RNA in our cells typically lasts minutes or hours, not years. It's chemically less stable than DNA, and not just by existing in only one strand. In chemistry, oxygen attacks. RNA has an extra oxygen that DNA has eliminated. That oxygen attacks, and it attacks a nearby RNA backbone, cleaving the molecule in two. RNA's very chemistry writes its own destruction.

And yet retroviruses write down their genes, the things they need to copy themselves, in this perfect impermanence. How queer then is this choice, if choosing can be used, to write oneself down in a molecule that can destroy itself. Viruses don't choose, per se, they simply evolve; an RNA virus, having that molecule chosen for it by luck, by accident, simply must make do, just like us living things.

The ways of a virus: reproduction, replication, mass production. A copy with no original.

This is the way a virus replicates: by sheer volume. This is the way it builds itself up: a capsid, the protein that holds a virus's shape, is made by repeating units of the same protein again and again and again until a 3D buckyball is made.

Looking at the shape of a capsid, made by mass production, its repeating units stitched together, an odd-shaped soccer ball, I cannot help but think of art. Copying something simple to make it something valuable was a powerful movement in pop art, which exploded in New York City in the '70s and '80s, just as HIV, a retrovirus, was invisibly and then visibly exploding as well. Andy Warhol's work directly addresses the postmodern concern with reproduction. From an object itself, through reproduction, sense and nonsense can be made.

A copy of a copy. A painting of a comic book hero. A screen print of an advertisement. Viruses can point to this as well, to the lies we tell ourselves about our collective values, who and what we consume, who and what we uphold and protect.

As Warhol painted his self-portrait, as Jean-Michel Basquiat painted his supermen, both were living in New York. Basquiat died in 1988. Warhol in 1987. They both died in the first plague years of a virus, HIV, but they both died of other things, unnatural causes. Drug overdose. The aftermath of an apparently minor surgery made complicated by poor health, healthcare avoidance, and having survived being shot.

What's the meaning of a reproduction? Warhol and Basquiat used reproductions to name (consumerism), to critique (one can only continue to consume), to render something violent (like war, via camouflage) visible and then re-create a new meaning from images that have become so commonplace as to be ignored.

What meaning does a virus have in its ability to replicate?

The virus has its meaning, its molecules, but without our body it is nothing. And then we can invent another layer, biomedicine, pills and vaccines that change the interaction between our body and a virus. The virus. Our body. Our biomedicine. Only through the three together can meaning be made; that meaning will shift over time as the virus, our bodies, and biomedicine do change.

First, the virus and our bodies. HIV is a retrovirus, its genetic material made of RNA, fragile to its own destruction. Of all our cells, viruses can only enter into the ones they can bind to. Each type of cell has a different suite of proteins. A virus needs something to hold on to. HIV, a retrovirus, binds to a protein, CD4, that's only on a specific cell in our immune system, a T cell.

Once inside, the retrovirus dumps its RNA into the soup of the cell. But its RNA won't last long. HIV has something almost unique in life: an enzyme that converts its RNA back into DNA, that stable molecule. Our cells don't do this, copy a temporary RNA molecule into DNA that will stick around. Information in cells flows only one way.

Viruses break every rule. A retrovirus is named for this fact of its life cycle: its enzyme is named reverse transcriptase, because information flows in the opposite direction of living cells.

Now that the viral RNA has become double-stranded viral DNA, just like ours, something magic, something dangerous, can occur. Like herpes, retroviruses will be with us for life. Not only will the viral DNA enter the nucleus, where our own DNA lives, but it will cut our DNA open and insert itself inside.

When we are infected with a retrovirus, its molecules become ours, inseparable for life. The DNA inside us bears no resemblance to the virus that once was; it doesn't have a membrane or a capsid, it cannot be seen even with the most powerful electron microscope. It's just DNA. Where is the virus then? It's nowhere, but in us still. A copy that renders the original not just invisible but forever gone. A copy that contains all the information needed to make the original again. Is it alive, then, copied and pasted into us because we're living? It gets copied with our cells. It's a virus without the image of a virus. Scientists call this a provirus. Just the genes, nothing more. But in that DNA is all the information to create the virus again, to build its capsid, to bud off from our cells, stealing our membrane.

Félix González-Torres, who died in 1996 of HIV, created art in almost every imaginable genre. But he's most famous for his installations: strings of lights that hang down a wall or stairwell, down the stairs to the coat check at the new Whitney Museum in New York; pieces of blank paper, a stack many feet tall, that guests are invited to take home with them; piles of candy, exactly the weight of his partner who died of AIDS, wrapped in bright, joyous, multicolored plastic, that onlookers can take and eat like a piece of communion bread, disappearing sweet on the tongue.

When a gallery installs an official Félix González-Torres, they don't get the lights or the paper or the candy. No. They get a piece of paper with instructions: Here's how you build the art.

They get the provirus; they make the object following the recipe they're given.

A provirus in Félix as he made his sculptures of light. That memory, that echo, that material information, that virus, a part of the body; the image of the virus, the memory of HIV, the material consequences of the disease, a part of our cultural body, too.

We've arrived, now, at a virus that has taken countless lives, except that we've tried to count them. How, then, does a virus kill? Again, infuriatingly, each virus kills in its own way, if it kills at all. And it depends on if we're talking about a cell that may die or an organism, a human, you, me. A virus may kill a cell in its replication: it takes the cell's machines and bursts the cell open to get out. But we have 43 trillion cells, and who'd miss just one? Most viruses that infect us may kill a few cells, that cellular death might itch or burn or inflame, but our whole bodies? Our whole bodies will live.

Everything depends on what cells the virus kills: HIV kills T cells, and without them, no immunity, and without immunity, no life. Influenza can infect many cells, and kill them, and damage tissue (lungs and heart and epithelium, our inside out) and so kill us, or give us a runny nose. COVID-19 kills many types of cells, just about any that has its right receptor, our own Ace2, as you will see. Almost any cell it can touch, it can kill, sore throat, broken heart, lungs—just one cell between the air and our blood—now permeable to anything, anything at all.

Viruses can bring death, we have to admit, even if most viruses, almost all, do not. Born in the 1980s, I know this. I've only ever been in a threesome: me and her and the imagining of a virus; me and him and the thought and fear of the virus.

CORONAVIRIDAE

Examples: Human coronavirus OC43 (HCoV-OC43; common cold, humans), HCoV-229E (common cold, humans; bats), bat RaTG13 (bats, asymptomatic), SARS-CoV-1 (SARS virus), MERS-CoV, SARS-CoV-2 (COVID-19)

I have a picture, hanging on my wall, of a good friend at a birthday party, a crown adorning his head, his eyes looking skyward, drunk on his own royalty. We can't see the shape of viruses using a light microscope. Their shape remains impossible to resolve. A stronger microscope that uses visible light won't work. The laws of physics prohibit it.

An electron microscope shoots—you guessed it—electrons. Light microscopes use light. Electrons are smaller than an atom. Using electrons, as opposed to visible light, we finally can gather information about the shapes of viruses. A friend of mine, in grad school, was using this technique to identify new phage, new viruses, from things like ocean water, just taking picture after picture and noting down the differences.

SARS-CoV-2 causes the disease COVID-19. SARS-CoV-2 is a virus.

SARS-CoV-2 is an enveloped single-stranded RNA virus, a coronavirus.

Coronaviruses have genomes made of RNA, not DNA. For RNA viruses, coronaviruses have large genomes. Very large. 30,000 letters or so. Big compared with HIV (under 10,000) and influenza viruses (around 14,000 across eight different molecules of RNA).

Viruses follow no rules except for the ones they write themselves. For many years, one of the only rules was that viruses are small. And then we found—we met—the mimiviruses. As big as a cell. A genome with the number of letters a bacterium might have. They'd been there all along, whether or not we'd discovered them.

Under an electron microscope, the coronavirus looks like a crown: a circle of membrane with spikes of protein sticking out all around. I glance up at the picture of my friend, at his crown, at the joy on his face, and yes, for that night, he was a queen.

Coronaviruses skip DNA altogether, using only RNA to both store and activate their information. Their RNA gets into the cell, a human cell, and immediately begins producing the viral proteins, using the machinery of our cell. Those proteins then copy the viral RNA not into DNA, but into more RNA, another machine that our cells don't have. We don't copy RNA into more RNA. RNA is only made from DNA as a template; the direction of information in life only flows one way, DNA to RNA. But viruses aren't living. Coronaviruses flow information from RNA to RNA to protein, and those proteins, then, can build a new virus. They never get into our nucleus. They never have to.

And remember that RNA is temporary. Once a coronavirus leaves our cell, leaves our body, its genes won't remain for good. To make us sick again, a new virus must once again not just arrive but infect. Viruses are limited, remember, by the way they store, the way they use, their information. If DNA is never made—and coronaviruses, like influenza, don't make it—the virus can't make a home in our home, our cells.

SARS-CoV-2 has killed millions already, a murderous, evil little thing. A menacing crown, as many crowns have been before it. How many monarchs were deadly in the history of humans? COVID-19's kingdom is our body, and it will rule: membrane covered in spikes that can stick to our cells and invade, breach the moats our cells build to keep out intruders, a Trojan horse that—as designed—can and will kill, our cells at least if not our whole body.

What meaning do we give to a virus?

SARS-CoV-2 does not have a crown. It has a membrane and proteins. Our cells don't have a moat and there's no Trojan horse under the microscope. The only crown it has is one we give it. And we're under no such obligation.

The meaning we give it is up to us. What is a soup can worth? A painting of a soup can?

Viruses do not want. They are not evil, they don't invade. They just are. They are a sack of membrane, they are the spike proteins, they are the RNA. They're an accident. They're the most abundant things on earth. No one said life made sense. It makes even less sense the closer you look. They have a capacity, one that can harm us, to bind to our cells, to replicate, and, in so doing, to kill us.

The meaning we give a virus affects how we live with it, live against it, die with it, die against it. The next time you hear of a virus, ask yourself what is its receptor, and so what cells can it bind to, me or a chicken or a bacterium? Lung or heart or brain? What is the virus made of, RNA or DNA, and, if invited inside, how long will it stay? Each virus, and the molecules it has, makes its own rules for not-living.

Most of the viruses we meet pass through us; many others die with us. If they want one thing, it's only this: to replicate. They are made to make more. Their desire is pure, singular. In a way, I envy them. In a way, they're just like you and me. Perhaps what we should be asking ourselves is how to be better than our basest biological need, better than replication alone. It's by living out this version of our humanity that we will best survive the rare virus that may mean to do us harm. Viruses have done us so much harm.

They've taken so many lives. But they are also an essential part of life; they were on this planet long before humans; they will remain long after we go. We live every day with the possibility of our own death. Its day will certainly come. It's in looking closely at life—and the nearly living things we're surrounded by—that we make it worth living.

3

On Going Viral

Which Viral Stories Do We Tell?

Start with a virus. In 2020, as economies worldwide shut down to limit the spread of COVID-19, we all stayed home and stared at our screens. We saw—on our screens—something happening to the sky, to the planet. Air pollution gone in Wuhan, China; water turbidity lower in Ahmedabad, India; mountains visible on the horizon from a smog-free Los Angeles in the United States. With people inside, animals returned: coyotes in San Francisco; ducks in Vegas; goats in Wales.

"We are the virus," came the memes on Twitter, "the earth is better without us." Nature is healing.

My response to this metaphor—"we are the virus"—is to ask what, in this imagining, is a virus and who, in this metaphor, are *we*. What stories does our culture tell about viruses, and how do these stories construct understanding of our bodies, our health, our neighbors, our lovers? In this viral story—*We Are the Virus*—humans have exploited the earth, emptying out and using up its resources, expanding and reproducing beyond the capacity of the earth, our host, and now the earth, our host, is slowly dying. Just like a virus, which uses our body, leaves us dying, emptied out of ourselves.

This is one viral story: consumption, illness, death.

And isn't this a true story? Just like Ebola or HIV might do, we're ruining the planet—our host—to replicate ourselves or our culture or to maximize our wealth. Others were quick to label this thinking ecofascist—using climate change or the earth's perceived needs as a way to eliminate "unneeded" human beings as if too many people were the problem. There's a pretty obvious way to fix the problem of too many humans and it's called genocide. But what's being produced isn't too many humans, it's too many humans with extreme, obscene wealth; maybe *they* are the virus, or capitalism itself. Couldn't *that* be true?

This viral story—even if it is true—is not without consequence. In this story of capitalism or racism as a virus, we are naming a virus as a thing of excessive self-interest, a thing that will overwhelm the whole, a thing that will kill, a thing *against* life itself. What would that mean for a person living with HIV or a person living with CMV or a person living with herpes, which is to say any person living on this living planet?

In this metaphor, there is a single viral story, and it is one of excess leading to death. This may indeed be one viral story. But applying it broadly to the larger cultural notion of not a virus but The Virus, we quickly find that the center will not hold.

All viral stories fall apart if we just consider them enough; there are viruses that consume and kill, but there are many, many more viruses that do not.

And you know by now that this has to do with the molecular life of the virus. There is no perfect virus; there are RNA viruses and DNA viruses; there are enveloped and nonenveloped viruses; and each virus has a different protein on its surface that allows it to kiss up on and interact with and maybe even get inside a different type of cell. Some viruses come for a few days, others for a lifetime. All of this is determined—in large part—by the molecular biology of the virus, a particular type of story, but one we too rarely tell.

"Of course," wrote Susan Sontag, "one cannot think without metaphors," which she defined, based on Aristotle, as "saying a thing is or is like something-it-is-not." "We are the virus" is therefore a metaphor, calling one thing another. Language is metaphorical in this way, where words stand in for the things they represent. "Table," the word, is not a table; the word conjures the table and, in a way, becomes it.

Microbiology is a real world, microbes are real, material things, but they are not like a table. When we say "bacterium" or when we say "virus," the word conjures the invisible material, things too small for *us* to see. Say "*Staphylococcus*" and "table" to a person on the street and ask what image each word conjures. A table is a goddamned table! If *Staph* means anything at all, it's a wound, an infection, not a bacterium. But *Staphylococcus aureus* lives on the skin or in the nose of between 30 and 50 percent of the healthy population without ever causing an infection, producing a wound. What is *Staph* then? The very language of microbes is built from metaphor; this looks like a crown, *corona*; this looks like an envelope, *env*; this looks like a spike, the spike protein now, *S*; these bacteria look like a bunch of grapes under a microscope, and so *staphylo-*, ancient Greek for that fruit still attached to its vine, nothing, or everything, to do with a wound, depending on how you look.

Yet, when it comes to health, Sontag demands that we remove metaphors from our thinking, our language, entirely. "But the metaphors cannot be distanced by just abstaining from them. They have to be exposed, criticized, belabored, used up." When looking at microbes, metaphors are so baked into our language that I think it is impossible to use them up. They will always exist; it becomes our imperative to *choose* metaphors and stories that have the capacity to not harm but heal.

HIV is a virus, a material thing. Functionally, while HIV might not be exactly the same sequence as it was in 1981, it's largely the same virus: the same genes in the same order that bind to the same cells that replicate in the same way and to the same effect. What HIV *means* now is not the same thing it meant in 1981 or 1987 or 1997 or 2007.

Viral stories will change, their centers will not hold.

One way to use up harmful metaphors of The Virus is to ask if they are true to begin with. But a true metaphor can still do harm. HIV does kill, but it is not the most typical virus. COVID-19 kills. Influenza killed my friend Sarah. But viruses exist all around us, always. Most of them do us no harm. To generalize, to make metaphors and narratives, from the *only* deadly exceptions as opposed to the benign rule is just one lazy way to not quite tell the full truth. If these stories—I will start telling some of them now—are almost-true but harmful, we must tell also the more common, or more careful, or more loving stories, the metaphors, that describe the world we live in *and* the one we want to build.

Lose someone to a virus. On February 11, 2009, while working in my lab studying phage, I got a phone call from a college acquaintance I didn't keep in good touch with. Worry, immediately. Why was she calling at 11 a.m. on a workday when we hadn't spoken in years? I stood up and walked to the lab's stockroom, the one place in the open-floor workspace with a hint of privacy that didn't also have a toilet in it. It was a small room, maybe 8 feet by 8 feet, with metal shelving units built against the wall, stacked with boxes full of petri dishes and pipette tips and filter units for sterilizing buffers. I took the call as I rushed into the room to make sure it wouldn't go to voicemail. I sat down on a large box shoved up against the shelves.

"Sarah died last night." She was crying.

I called Andrei immediately. He worked just one floor below. "Come to the stockroom. The Darst lab stockroom." In 2 minutes he was there, hugging me.

"Honey, what happened?" It was the first death of a close friend. How could I even explain?

Sarah Tilman was one of my best friends from college. Only weeks before, I'd taken a cheap bus down to Washington, DC, to visit her. I stayed on her couch. Both biology majors with French minors, she now taught bio at a public high school and I was starting my PhD, a couple years in. We'd been study buddies but quickly became friends. I had a crush on her, of course, but it never amounted to much. She was from Minneapolis; her dad was an ecology professor. She'd sold her car because, after moving from Minnesota, she found she didn't need it out east. She taught high school because she loved kids; she was part of the Washington DC Teacher Corps because she thought Teach For America sucked—they just have a bunch of idiot kids working there, they teach for two years, all first-year teachers suck, and then they go off to work in policy or whatever. Sarah was in it for the long haul, 10 years at least, because she became better and could give more to her kids.

Those are facts about my friend, who died. I'd tell you, but I forgot her favorite color. Excuse me a second. I have to get a glass of water.

Two months later, I sat across the table from a lawyer in the Starbucks near my lab. She asked questions about Sarah. Was she thinking about grad school? Yes. How long did she plan to continue teaching? At least two or three more years, she was deciding between grad school in education and in biology. Sarah shouldn't have died. This lawyer, hair tied up, dressed in a dark gray suit, was here to determine how much her life was worth. I hated her, but smiled at her and pretended everything was fine. Just a normal meeting in a normal Starbucks on a normal Wednesday afternoon. When I'd visited Sarah just weeks before she died, we'd had a big debrief about her next steps. This lawyer needed to know that information so she could plug it into a formula about future earnings and decide how much the doctors who led to her death needed to pay her family. I saw the spreadsheet in her head, the numbers ticking off, what my friend had become: money.

Sarah died of flu. Two weeks before, we went out drinking. As a teacher, she was worried about the winter virus season, so we used Emergen-C powder as a mixer, just powder and vodka and ice. We knew to mix the powder in first before cooling the drink on ice because the powder would be less soluble in cold water; we'd learned that together. We were with another friend—I don't remember, now, who. The three of us, 27 years old, children really, drinking vodka and Emergen-C before going out dancing at a gay bar.

She died of a virus. Influenza A. The virus had infected cells in her heart, which it occasionally does, but the hospital had failed to recognize a severe dip in her blood pressure. They sent her home. Hours later, feeling ill again, she called an ambulance. When she stood up to walk out of her home, her blood pressure dangerously low, she had a heart attack. They never could restart her virus-laden heart. She died.

In a classroom, in 2004, I'd sat next to Sarah and learned about influenza, how it's a virus with eight separate RNA molecules, viral chromosomes in a way. We learned, too, about how these eight molecules and the fact that flu infected swine and birds and us and could swap these eight molecules around made it a high-risk virus for becoming a respiratory pandemic, killing millions, just like in 1918.

It didn't take a pandemic for Sarah to die. She died a viral death. My grief was too much to bear.

At her funeral, everyone spoke of heaven and seeing her again, but I didn't believe in that, and I don't think Sarah did either. I remember thinking to myself that it had never been as hard to not believe as it was just then. I remember thinking you can't just believe when you want to, you had to believe all the time, and I knew I wasn't capable of that. I'd tried. I was sitting in a church crying. She had an open casket but I couldn't look at her body, dead. I knew my goodbye to her was final. But looking back now, I feel like I say goodbye each time I remember her, and then remember, again, that she's not here.

In my email, from the lawyer, with the subject line "Tilman case":

> I just wanted to let you know that this case settled on Friday, so there will be no trial appearance asked of you. I think that —— and —— Tilman were pleased overall with how things went, and felt a sense of closure. They are wonderful people. Thank you very much for your assistance, and your willingness to meet with me and talk about Sarah, it was definitely helpful.
>
> Wishing you all the best in the future!
>
> Regards,
>
> ——

I know what viruses can do. A virus killed one of my best friends. But to stay mad at the virus is to stay mad at the world. She died because doctors didn't care for her well enough. Care could have saved her. It should have. It is care we need in this viral world.

I think I'm writing so much about the lawyer because if I even tried to describe what she looked like, Sarah, I would remember that I'd never see her again. The lawyer was tall and thin with dark hair; she had thin eyebrows and kind eyes, which—I think—she practiced. Sarah was short and rail thin, she had brown hair that she sometimes straightened but that naturally curled. Her smile, when she was smiling, crept up one side of her face more than the other, just slightly, which is how you knew she hadn't practiced it, it was real, and there she was, smiling at you.

Conjure a virus. Just years before we first saw a virus on an electron microscope, author Zora Neal Hurston wrote one down. Janie, somewhere near the end of *Their Eyes Were Watching God*, having left one husband and buried another, is something-like-happy with her third husband, Tea Cake. They live out on the muck farming when a storm hits, a hurricane, and everything is mud and everything else is water. In the chaos of the storm, Tea Cake is bitten by a rabid dog. When the storm finally goes, rabies makes the chaos of the storm physical in the body of Tea Cake:

"Tea Cake was lying with his eyes closed and Janie hoped he was asleep. He wasn't. A great fear had took hold of him. What was this thing that set his brains afire and grabbed at his throat with iron fingers? Where did it come from and why did it hang around him?"

"Tea Cake had two bad attacks that night. Janie saw a changing look come in his face. Tea Cake was gone. Something else was looking out his face."

Rabies, a virus, is taking Tea Cake away from Janie. But it's not just killing him. Before he dies, the rabies is turning him into something he isn't. Even while he's alive, he's gone, his body no longer under his control.

Rabies is a single-stranded RNA virus that infects muscle cells and neurons, the cells used to send information in, to, and from our brains. This is why rabies take us away before we die: once it arrives in our brains, it kills our neurons. And which neurons it kills might not be random either; evolution appears to have favored rabies viruses that specifically kill regions of the brain resulting in aggressive behavior and fear of nothing except water. It's from the salivary glands that the virus must find its next host, through a bite. Aggressive behavior and not swallowing anything, not even spit, probably helps the virus move on.

Janie is protective of Tea Cake, even in this state: "Folks would do such mean things to her Tea Cake if they saw him in such a fix.

Treat Tea Cake like some mad dog when nobody in the world had more kindness about them."

But the virus had its own story to write, written there in its RNA genome: "Tea Cake couldn't come back to himself until he had got rid of that mad dog that was in him and he couldn't get rid of the dog and live. He had to die to get rid of the dog."

These days, and even then, rabies doesn't have to kill. It's not the only version of this story. The doctor that Janie sees after the hurricane tells her that another version of everything was possible: "Some shots right after it happened would have fixed him right up," he says. Rabies rarely kills now; we can vaccinate after an animal bite right up until the moment when the virus reaches the brain, where it will do its worst damage. Rabies is still out there, in the world, and it probably always will be because of its presence and spread in animals. But through science and medicine, we can insist that it doesn't have the final say.

This is one viral fear that we have, being taken over, losing ourselves, effectively becoming a *zombie*. Something in us eating our brain but letting us live. Something taking over our brain and telling us to *chase! run! bite!* but absolutely not, ever, to drink. "And she was beginning to feel fear," Hurston wrote of Janie, "of this strange thing in Tea Cake's body." This *strange thing*, this nightmare, this virus, turning us into something like itself: not living but not yet dead.

Imagine going viral. See, viruses can mean so many things, and here the excess of the virus is not to be feared but longed for. Susan Sontag wrote that "one set of messages of the society we live in is: Consume. Grow. Do what you want. Amuse yourselves." In late capitalism, we exchange our labor for money, and that money for things and experiences. "The very working of this economic system," she wrote, "which has bestowed these unprecedented liberties, most cherished in the form of physical mobility and material prosperity, depends on encouraging people to defy limits."

This desire she names for growth, consumption, for never-enough, is central to the viral metaphors we use. Some viruses have the ability to replicate within hours, infecting one cell, then a hundred, then a thousand, so difficult for the fully living to achieve. This exponential growth, too, is seen in the ability of information to flow via the Internet: if one person shares a video, and then three people share that video, and then each one of those people who receive it has three people share that video in turn, you can reach an unprecedented number of eyes and minds in an astonishingly short time.

The use of viral metaphors to describe information flow through culture likely begins, in the modern era, with Richard Dawkins defining a *meme* in his book *The Selfish Gene*. Dawkins's project is both revolutionary and deeply troubling; he is applying the laws of Darwin (survival of the fittest) not to the level of individual organisms like you or me, but to our genetic units, to the genes we carry. It's our genes that "want" to survive and multiply. We, as living, thinking people, are helpless carriers of the true tyrants of evolutionary need: our DNA. That Dawkins would end up being so shitty and racist is not surprising to me; social Darwinism is an extension of Darwin's theory of survival of the fittest into social structures: poor people are less good and deserve to die. Dawkins applies survival of the fittest to the most microscopic level, social

Darwinists to the most macro biological level: culture. Both overextend the metaphor beyond the possibility of something-like truth and into the realm of overt, measurable, too-flat-to-be-true harm.

Dawkins's ideas do hold some water in some cases (sequences related to viruses are the most likely to evolve "selfishly," as he argues). All these cis white men! In their imaginations, the possibility that life could be collective, communal, isn't real. The metaphor is always individualistic, always selfish, always a brutal war for limited resources, the options being death alone or survival alone. Of course, there are many ways in which evolution does indeed act on organisms (and even communities) as opposed to genes alone. In his work, though, he defines a meme: a piece of information that flows through culture. Ideas or performances can evolve over time a lot like life: replication and mutation, competition and variation. And the best memes will travel the farthest the fastest, through competition, the survival of the clickiest (sorry).

Viruses—and going viral—promise us just exactly what COVID-19 demonstrated in early 2019: exponential growth seemingly without limit. Living organisms follow rules of replication and reproduction, where one becomes two becomes four becomes eight; viruses know no such laws.

On Twitter, one can have only 2,000 (or even 200) followers and write that perfect tweet. A writer I used to date told me, one night, that he knew for a fact that people in the Kardashian family read his Twitter feed, and one day, maybe, they'd re-tweet him, and from there . . . I didn't get a clear sense of what he thought would happen next.

Leave Britney Alone! and Charley bit my finger! and David after Dentist! and hash tag FreeBritney! and Antoine Dodson! Dodson went viral for a local news clip recorded the day after his sister was nearly raped in the home they shared. It was full of Black queer anger and resilience. It went viral. Within weeks, he had a song on iTunes, appearances on NBC, a merchandise line, and a website. He endorsed a sex offender tracker app. His goal was to raise enough money to get his family a home in a safer neighborhood. Within a few months, his family was able to move. Within two years, he was working on a *Beverly Hillbillies*–type reality show about his family's attempted move to Los Angeles, which never materialized. In 2018, he posted a YouTube video using catchphrases from his video to promote his business in real estate. CelebrityNetWorth.com, in 2020, listed his net worth as $100,000, which is less than half the median home cost in Huntsville, Alabama, where he now lives.

Of course there are exceptions, artists savvy and talented enough to use virality to create a lasting creative platform and not just 15 nanoseconds of fame. Lil Nas X is my favorite; his TikToks are hilarious, his music catchy, his celebrity joyous. His trolling of folks who told him he was going to hell with his "Montero" video? I mean, we don't deserve him.

Lil Nas X, though, is the anomaly. That so many stories of viral success end with a family-sit-com return to life-as-before doesn't seem to deter our collective desire for this type of attention. I work on it constantly, but I feel it in myself, too. We may want to go viral for the right reasons—for an essay, not a banana peel nosedive on camera—but we want to go viral nonetheless. The viral story: break the rules of life, spread everywhere faster than we can even imagine. For so many of us, like Dodson, like many folks I grew up with back home, the options for escaping one's hellish predicament seem so limited; going viral might be a lottery you know you're certain to lose sooner or later, but at least you're holding, in your hand, a piece of paper that could turn itself into a ticket out.

Going viral also allows the possibility of building one's celebrity (and therefore earning ability) outside of traditional media. One doesn't need to be the child of an actor to have access to media industry insiders—agents and casting agents. One doesn't have to be cast in a VH1 reality show. One just needs an Instagram or Twitter or Twitch or TikTok or whatever will come next. The raced, classed, gendered aspects of this are obvious. Many minoritized people have gone viral, and some have used that virality to build a lasting career. Writer Kiese Laymon dedicated his first book to "the Internet" after a viral essay on Gawker.com helped him sell thousands of copies with no support from his small press. He'd spent years being ignored by mainstream publishing who didn't think his stories would sell to their imagined readers. The Internet—virality—showed that he had readers, that he could sell books to people publishing didn't think *read.* With virality offering novel pathways to creative success, especially to queer people and people of color, no wonder it's also derided, mocked, called immature, pathological, superficial.

If going viral (or playing a professional sport, or taking on hundreds of thousands of dollars of college debt) is needed to be able to have the resources to live a good life, there is a problem not with our desire for virality, but in the circumstances of our everyday lives, of the cultural body we make up. What's wrong with our everyday lives to create this need? We too often lack resources, financial stability, healthcare independent of our work, the right to work, the right, too, to leisure and making and consuming art. Kiese Laymon calls this having healthy choices and second chances, and we all deserve both. Who has access to either depends on gender and sexuality and race and class and geography and so much more. Here again the metaphor of the virus is pointing to the wrong harm. Going viral isn't pathological, nor is our desire for virality. Our culture—where so many live so close to death on a daily basis—profoundly is.

And yet this desire, this obsession, can drive us mad, even as we blame ourselves for not surviving well enough in a death-making world.

Watch a virus. In the popular (but not very good) American film *I Am Legend*, Will Smith plays the sexiest ever virologist living in the aftermath of a deadly viral pandemic. In his basement lab, he searches for a cure. He does clinical trials on lab rats. On the radio, he calls out for other survivors. At night the zombies come out, called Darkseekers, to hunt for uninfected flesh. The virus, originally rolled out as a vaccine-like cure for cancer, killed most of humanity and turned most of the rest against itself. This, like in most zombie movies, is a death worse than death, a neverending death, a death wherein we feast on the living, whether we once loved them or not. This viral story, this metaphor (the virus turning us into something we're not, the virus turning us into an image of itself, always hungry for a new host) is an extension of the rabies viral stories like Hurston's.

The film also uses one of the most common viral stories; in a flashback to the initial viral crisis, with a cancer vaccine mutating and killing those never vaccinated, Will Smith's wife asks him, "Did it jump? Is it airborne?" Once the virus is airborne, invisible but everywhere, nothing can stop it. Never mind that many viruses already are airborne, or travel by air, including rhinoviruses, measles, influenza, and yes, coronaviruses too. "Did it jump?" Can the virus jump from body to body through the air, no physical connection? What then? What now?

I'm a virologist (sort of) and I have a dog (unlike Smith's German shepherd, Max is 15 pounds and afraid of plastic bags blowing in the wind). In the world of *I Am Legend*, as in the postapocalypse of *The Road* or *The Walking Dead*, I would have been sure not to survive the mass death, even if by my own hand. I'm not the type of virologist who would watch his family die and stick around looking for the cure. I would just choose to die, too.

In this story, in this viral metaphor, the virus itself stands in for the hubris of humanity. It comes from an attempt to subvert nature, to cure cancer by using a virus. The virus is there to show us nature will always win; it's human-made but—like nature itself—cannot be controlled. "God didn't do this!" Will Smith screams at the first human he meets in years. "We did!"

For centuries, syphilis did what we now imagine rabies to do. Because tertiary syphilis infects and kills cells in the human brain, it can affect behavior and mood. It can turn us into a living non-self. In the twentieth century, antibiotics intervened. Then viruses filled that void. From Hurston's very real rabies to Will Smith's imagined Krippin virus, the human horror is becoming a shell, a body without a brain. Viruses are our tool to tell stories of this horror.

I watched my grandfather die, a shell of himself, barely able to speak, not able to walk. He couldn't remember who I was. Why don't we tell of this essential human fear—losing our minds but not our bodies—with the stories we have, the true ones, about the people who we've lost this way? For one thing, unlike Will Smith, my grandpa couldn't take off his shirt and do L-sit pull-ups on camera. Dementia doesn't make you chase down folks, although it can make you bite. It may be partially genetic, but it isn't infectious—unless you mean its mad cow cousin. Viruses have become a way of telling the story of lost cognitive ability while actually looking away from the vast majority of lives we lose this way. We seem not to deem most of the people dementia happens to worthy of telling stories about. No heroes to be found there. No saving the world, no fixing it, no six-packed scientists, fighting the very thing that might kill them.

In the original book *I Am Legend* by Richard Matheson, the zombies are not made by viral infection but by infection with a bacterium. In the years between the book (1954) and the movie (2007), we witnessed HIV and Ebola and SARS. Rabies, the actual virus that makes us zombie-like for a time before we die, has faded from our collective consciousness. Here in America, it is extremely rare in humans. The real virus has been replaced by a metaphorical one, a virus that can do anything, the idea of a virus. Even in that horrible year, 2020, what we imagined was worse than the worst nature can do.

Smell the virus. I know I did. I read *The Hot Zone* for the first time around 1996. It was after the paperback came out, and I remember the feel of it in my hands. A precocious tween, a voracious reader, I was more curious than horrified. The book tells the story of virus hunters who, when a new disease emerges in humans, travel to the site and try to find a pathogen, a virus, to explain it. There I was on the couch in rural Washington, but really in a bat cave in central Africa. There I was riding the bus to school as the sun came up, but really I was in a monkey facility outside Washington, DC.

The Hot Zone details researchers working on the biosafety level 4 hemorrhagic fever viruses, including Ebola. BSL-4 labs do everything possible to contain the most deadly viruses we know—scientists don positive-pressure suits, airlocks protect exits, everything inside is decontaminated before leaving, the people included. Biosafety levels range from BSL-1 (working with harmless organisms like nonpathogenic *E. coli*) to BSL-4, where scientists study live patient samples of the world's deadliest pathogens. Hemorrhagic fever viruses can kill between 50 and 90 percent of those who are infected, and the deaths can be quite gruesome and bloody.

In the 1990s, *The Hot Zone* was followed by the hit film *Outbreak*, starring Dustin Hoffman and Rene Russo as two (very attractive) virologists. The film opens with a quote from microbiologist Joshua Lederberg: "The single biggest threat to man's continued dominance on the planet is the virus." Joshua Lederberg's wife—Dr. Esther Lederberg—discovered phage lambda while they were both at the University of Wisconsin, talking viruses—probably—from coffee in the morning until cocktails before bed. But lambda comes later.

Not surprisingly, what follows in the film *Outbreak*—and what we saw in *I Am Legend* as well—is a battle waged between a virus and military scientists. It is true that a significant portion of our country's pandemic response depends on the national security and military apparatus, and the military does employ expert virologists and vaccine scientists. This placement within our government, and Lederberg's quote, tell us that much: man is meant to dominate, the virus is a threat.

Joshua Lederberg certainly had cytomegalovirus in him when he spoke of a viral threat, almost certainly had at least one herpes virus in him. He was made—then—of endogenous retroviruses; he—like all humans—has known viruses since the moment of his introduction into the earth's atmosphere, his first breath and cry. What viral stories do we tell, and how do they fit into our preexisting narrative structures of war, of domination, of colonialism, of all the death making we ourselves have invented?

Outbreak and *The Hot Zone* both tell a story of a virus that is a mutation or a mistake away from ending humanity. In *The Hot Zone*, we see a virus arrive on our shores, unnoticed, in a monkey, only miles away from the US Capitol and the White House. In *Outbreak*, the virus arrives and then mutates. Instead of traveling only via infected fluids (like Ebola and other hemorrhagic fevers), there's a new spike visible under a microscope. Dustin Hoffman, after looking up at a vent in the ceiling of the hospital, from inside a yellow BSL-4 suit, looks directly into the camera and says, quietly but with urgency, "It's airborne." The virus has mutated, we cannot see it, it travels by air, it kills in only days, it liquefies our insides (a woman spitting up blood into a bin is visible in this very scene with Hoffman), it's nature unleashed, it's from the Congo, it's from the jungle, it's everything we're afraid of, it's here, already here, and moving.

This viral moment for Ebola and hemorrhagic fevers came in the mid-1990s in the height of the HIV pandemic. Both HIV and Ebola emerged in central Africa, and the *Heart of Darkness* narratives are evident in the stories we tell about both families of viruses. Viruses spill over from animals into humans all the time and all over the world—coronaviruses in the Middle East and China, H1N1 influenza recombined from pigs in North America in 2009, hantavirus from deer mice in the American Southwest. The stories get told through colonial lenses: viruses, close to nature, in unspoiled dense jungle, passing first into Black people in Africa, then arriving, finally, on *our* shores. Susan Sontag reminds us that "there is a link between imagining disease and imagining foreignness." The truth of viral spillover is that it can and does happen anywhere animals and humans meet, which is anywhere, including here. The truth of the stories we tell about it is that it happens *there* and arrives here, *The Hot Zone*, "the China virus." This is a violent metaphor, the foreign virus, one that leads to xenophobic attacks like we saw on Asian immigrants in response to COVID-19 and Haitian immigrants in response to HIV.

So it should be of no surprise that the machinery of American colonialism—the military—would show up in our viral stories. As a COVID-19 activist, I met (via Zoom) with vaccine researchers in the army. The American military has an interest in keeping their own soldiers safe, which is why research into diseases like malaria continued in the Defense Department even for the decades where this work was halted nearly everywhere else in the country because, simply put, Americans didn't often die of the disease.

Historian Jim Downs wrote about a smallpox epidemic among formerly enslaved people in the American South during the reconstruction era. For six years, from 1862 through 1868, thousands of Black Americans died. The epidemic was more deadly than other smallpox outbreaks at the time due—Downs argues—to a slow response and ineffective quarantines, poor quality or no available healthcare, and indifference at all levels of government and in the media in both the North and South. It was the military that responded to the outbreak with attempts to quarantine and once again the military who saw the smallpox epidemic as a "natural outcome" of freeing people who weren't naturally fit to be free.

Whiteness, as ever, the freedom to harm, the ability to take life. The military, as ever, one mechanism of that harm's doing.

The placement of this essential research on infectious diseases, from malaria to smallpox, within the Department of Defense shows us how we are willing to invest to protect our *national* body from threat. Malaria kills only abroad; until the Gates Foundation started funding malaria research in the global West, the researchers studying the disease were almost exclusively working in military labs, ostensibly out of interest for our soldiers, our national body at risk. Colonialism deprived so many of the world's countries from developing their own scientific infrastructure to study the diseases that impact them directly. The millions around the globe who die of malaria every year were deemed less worthy of protection than our American national body and the soldiers who are its physical incarnation.

Given our current national governance, the military is the only way to respond as quickly and forcefully as needed to a crisis. And why? The military has the funds. Money is essentially unlimited. War is a place, perhaps the only place, where questions of practicality fall away.

These army labs are dedicated not to killing others, but to saving lives. I cannot look away from the irony. But it doesn't have to be this way. We could invest in saving lives in the same way we currently do in "national defense." We could reimagine national defense not as something to defend our borders or our national body, but as a way to insist on the longest and best life for everyone in our nation. Not war, but care. Not geographical borders, but the borders of our own bodies, our skin and guts and lungs. How much money is too much to spend in an emergency? The limit does not exist. What is an emergency? Yes, a pandemic, but yes, too, a friend's cancer diagnosis, a good friend's influenza.

In my imagined pandemic movie, in some alternative version of this earth, an unknown virus shows up, and it can kill, and the researchers in their BSL-4 suits who arrive don't work for the army, but for a National Center of Wellness, an office dedicated to the security of the bodies we all hold most dear.

The way Richard Preston tells it in *The Hot Zone*, the Ebola virus liquefies our insides before we die, as we shit and vomit out our organs. This is a viral story, a commonly told one, the thing that comes on so quickly and digests our flesh even before we die. An overtaking. A bloodbath. The thing inside consuming us whole. Nothing left in the end but a puddle. Nothing left of us, nothing safe to bury. While I was reading one particularly gruesome chapter on the bus home, a queasy kid two seats behind me did puke. I didn't hear anything—my mind was entirely in the Congo—but the smell. The scene as I remember it now: an Ebola patient vomiting a mix of darkness and blood. The world of the book came up around me even more as the smell became my only sensory experience, the only thing besides the book in my world. I put my forehead against the cool glass of the bus window and I saw the blue sky, the last thing I saw before retching up my own guts and looking down to see how much blood was in my vomit and wondering how many hours I had left to live.

Use a virus. In the late eighteenth century, hundreds of years before any human saw a virus under an electron microscope, French novelist Pierre Choderlos de Laclos did just that. His epistolary novel *Les Liasons Dangereuse* became, in 1999, a favorite movie of mine—*Cruel Intentions*—that made me feel things I didn't yet understand. In the original, the Marquise de Merteuil and the Vicomte de Valmont scheme mostly, it seems, to entertain themselves. Reading the original, in college, I learned the word *libertine*, one who lives against the mores of one's moment in time, flaunting the accepted moral code.

Merteuil plans for Valmont to seduce Cécile, who's in love with Danceny, but Valmont plans to seduce Tourvel, whom he eventually comes to love, genuinely love. Valmont does seduce Cécile for revenge, which Merteuil then uses to get her own revenge, leaking news of the affair to Danceny. Danceny kills Valmont in a duel, but before he dies, Valmont gives Danceny the letters proving Merteuil's schemes. Shamed, Merteuil flees.

"Madame de Merteuil fell ill," Laclos writes, "with a very high fever which was at first thought to be caused by the savage treatment she had received; but last evening we heard that it was smallpox, of the confluent variety, and extremely virulent. It would in fact, I think, be a blessing for her to die from it."

A virus so brutal, so painful, that death is preferable.

But no, viruses don't always kill. Days later, "Madame de Merteuil's fate seems finally to have been sealed and in such a way that her worst enemies are hesitating between rightful indignation and pity. I was surely correct when I said it would be a blessing for her to die of smallpox. It is true that she has recovered but she is terribly disfigured; in particular, she has lost one eye. As you may imagine, I haven't seen her again but I have been told that she is absolutely hideous."

"Her illness," Laclos writes, "had turned her inside out and that now her soul was showing on her face."

A virus that doesn't kill, but renders ugly, bringing the body in line with one's inner self. There would be no happy *libertine* ending for Valmont or Merteuil (or Tourvel or Cécile either, of course). For Laclos, smallpox brings a final justice. Merteuil, the scheming woman, deserves to suffer, and what better way for a woman to suffer than being stripped of beauty, and so power. For men, death. For women, life without beauty.

A virus can be anything, even—in the hands of Laclos—a perverted form of justice. There's no viral center, not anymore, to hold on to.

We end this essay where we started: with a virus, but one that only infects bacteria. So, we end with a phage, not *Myoviridae* but *Siphoviridae*, its cousin. Esther Lederberg, at the University of Wisconsin, discovered lambda phage in 1950, when these viruses were commonly used as basic tools to understand bacterial genetics. We'd only just shown, conclusively, that DNA is the molecule that confers heredity (and not, as many believed, protein), but the double-stranded nature of DNA wasn't yet known. Phage were simple things that could be grown in a lab, and their activity (killing bacteria) could be easily measured: mix virus and bacteria together, then count the number of plaques where the bacteria have died. More plaques indicates more or better virus. We call these phage that can kill lytic for their ability to lyse—or kill—cells. The word *lyse* itself is from the Greek *luein*, "to loosen," and this is what cells look like when they lyse: they burst loose. Lytic phage are viruses that burst cells open, spilling new copies of themselves out, one common enough viral story. Lytic viruses are viruses that kill, at least at the level of our cells.

Phage lambda tells a viral tale. Its host is *E. coli* bacteria, and it binds to the surface of its host cell, where its DNA will make its way inside. Once inside, the viral DNA is recognized by the bacteria's machinery to turn its own viral genes on. A burst of gene expression then, viral genes at higher and higher levels, eventually among the most common things in the cell.

The virus now has the enzymes it needs to copy its own DNA, and all the proteins it needs to build itself back up. And in a wave these proteins are made, and in a wave its own DNA is copied, and the bacterial cell itself is struggling to survive, all its resources now a factory for the virus, not for itself. It's out of its own control. It isn't feeling itself. Zombie bacteria, doing the virus's bidding, poor thing. If only we could shoot it now, its fate already sealed.

Because yes, the virus is in control now. And once its DNA is copied and once its proteins are made, it will self-assemble, build itself up, and it will pierce the membranes of the bacterial cell, its host, leaving the bacterial cell itself dead, used up, and releasing more viruses—hundreds, thousands more phage that can go out into the world and meet other cells, other bacteria, they can bind to, kiss up on, eat alive, and kill.

Write another viral story. Same virus: phage lambda. The story starts the same way, too. A virus meets its mate, its host cell, once again an *E. coli* bacterial cell. Like before, it pushes its DNA inside, and its DNA, once inside, is recognized by the *E. coli* machinery, and viral genes are turned on, made by the cell, viral proteins now swimming in the soup of the cell, in its guts.

Something different happens now. One protein, CII, gets made just enough. CII is an inhibitor; the virus creates a brake for its own development. If enough CII gets made, and sticks around, just fast enough, it will block the virus from continuing down its path of destruction, of death.

Instead, the virus makes a different set of genes. These genes do something odd, something—over the long course of evolution, the long history of the virus and its host—that must be strategic because it still occurs. The DNA from the virus doesn't lead to the cell's destruction. The viral DNA instead gets cut into the DNA of the bacteria, the virus and the host now becoming one large thing, one super molecule.

Every time the bacterial cell divides, the virus gets copied too. Every cell made from the bacteria will hold the virus in it. The interests, now, of the virus and host merge into one. The better the bacteria does, so, too, does the virus.

So we have seen two viral lifestyles: lytic is the deadly pathway; lysogenic is the silent one.

Phage lambda's lysogenic pathway is how it got found. Esther Lederberg was stressing *E. coli* with UV light, which we know introduces mutations. Perfectly normal *E. coli* cells, she thought. She wasn't looking for a virus, and yet, when she stressed the cells with UV light, she found one anyway. Lambda can reactivate, become lytic again, and lyse and kill bacterial cells when the bacteria itself is stressed. The thinking is that if the bacterial cell is going to die anyway, it's time for the virus to get out. A certain number of her cells, when stressed, had their lambda phage activate, and the virus made itself again and killed that *E. coli* cell before spreading to the cells around it, killing them too.

Lysogenic, from the same latin root for "lysis, cell death"—but now with the suffix *genic*, for "production." Lysogenic, production of death. Lysogenic phage were identified when they once again became lytic. It was then so much easier to measure death, visible to the naked eye on a plate, than when the virus was silent, the genes for the virus passing themselves along for generations.

Even in silence, viruses are named for death. We did this to them. We named them this way.

This virus can do both: it can kill, and it can be silent. But the virus is so simple. The choice is made by an odd mix of strategy and luck. I'm bigger than a virus. I have no ability to control luck, but I can decide, myself, what I want for a strategy, and to build that around care, not competition.

We have so little control, and I, at least, take solace in this. When life seems impossible, when you wake up and stub your toe and you're hungover even though you only had two beers the night before because you're almost 40 and your body does that now, for some reason, and you have a day full of meetings that don't matter and your best friend is mad at you for something that deserves it, think about CII. Somewhere, in someone's gut, even now, lambda is meeting its host cell. What will it do? Maybe the *E. coli* cell hasn't long to live, but maybe—too—CII will save it, and the two things will become one, their fates now forever tied in a molecule, DNA, at the center, as always, of life and of death.

There's no such thing as *the* virus. Ebola is a virus that can be so deadly and replicate so rapidly that a human body goes from fine to fever to leaking our liver through our anus in 24 hours. HIV tells a different story: invisible, lying in wait inside us for years, just waiting to make us die slowly, painfully, outwardly shamed by the sex we had had.

Both these stories are partly true. It's where the stories come together into one thing—the virus—that they fall apart entirely. What does a virus mean to us? That one thing could contain two drastically different nightmares in its 120-nanometer shell shows just how strong weak metaphors are. These metaphors (from *The Hot Zone* to art from the "HIV plague years," as if they aren't still here) had lasting impact on me and so many others not because they were scientifically accurate or because they best represented what a virus does or can do, but because they were horrifying. Horror grabs our attention, even when it's a lie, even when it's a metaphor.

The problem wasn't illness. The problem never is. Illness is a fact of life. The problem is our inability to provide care to all.

It's not that these metaphors of the virus are untrue. Rabies infection does damage the cells of our brain and make us unrecognizable before it makes us dead. Ebola can come on so fast and kill so quickly. HIV can lie dormant for a decade before the onset of symptoms. HIV, COVID-19, Ebola, and rabies can kill.

But each of these viruses acts so differently, each virus tells its own story. And even then, the better question is which viral stories get told to begin with. There are so many viruses on this planet, more than people, more than bacteria, even. They're more common than any living thing. The only stuff on earth more common than viruses are abiotic things: fully dead, not even halfway living.

And most viruses do nothing. How boring, how painfully banal.

Who'd want to tell that story, and if someone were dumb enough to do it, who would have the patience to read it all?

We refuse this metaphor of the virus at our own peril. Telling only horror stories makes us imagine only horror. Most horror stories involve everything but a virus. We have shown ourselves to be perfectly capable of actively or passively killing others without the need for a pathogen at all.

We must recover the notion of the virus and stop letting viral stories be told only through the exceptions to the rule. If we were to look at all the viruses on the planet, right now, how many of them could kill a human? It's a rounding error larger than zero. This is not to minimize the lives lost to HIV or COVID-19 or to any other virus. As long as we live, viruses will almost-live alongside us. That's the only viral metaphor we can apply to their entire kingdom. We live on a viral planet. They were here first. We are their guests, not hosts. That is a viral story worth telling.

Viruses may terrify us. We tell stories of a happy world punctured by a viral pandemic, leading to death or zombification. Will Smith watches as his wife and children die. The Reston virus kills monkeys just outside Washington, DC. Gwyneth Paltrow dies in the first 5 minutes of *Contagion*. But even these stories may not be useful ones. They didn't—it would seem—lead us to prepare for the reality of a very real viral pandemic of a respiratory disease, one all the more nightmarish for the ways in which it became entirely mundane to bury 3,000 Americans a day.

In *Their Eyes Were Watching God*, Tea Cake dies from a very real virus. But Janie's reaction reminds us that love and memory outlast even death. Tea Cake's rabies shows us that life is always precious, that bad luck can come at any time. Tea Cake dies saving Janie, but not in that moment. He was bitten by the rabid dog then, but rabies takes weeks to travel from the muscle near the bite to the brain. Tea Cake's loss doesn't upend the world, it doesn't kill most people on earth, just the person most dear to Janie. That's how viruses do work, in the world, when they do kill.

After she shoots not-Tea-Cake dead to save herself from dying, Janie gets a vision of him: "Then Tea Cake came prancing around her where she was and the song of the sigh flew out of the window and lit in the top of the pine trees. Tea Cake, with the sun for a shawl. Of course he wasn't dead. He could never be dead until she herself had finished feeling and thinking."

Tea Cake, the real Tea Cake, is in her, a memory.

The one time that I longed to get HIV, it was to have a connection to a boyfriend who wouldn't publicly acknowledge our relationship; the virus would be a memory of it, in me, a DNA sequence present, incontrovertible, shared between us, as long as we both shall live. He couldn't deny that. If I couldn't find a man to marry me, at least maybe I could find one to be with me forever in that way. I've learned since that I don't need that DNA: my memory of him, of all of him, is enough.

In the last season of NBC's *The Office*, thinking back on their time working together, Jim tells Dwight, "Every time I've been faced with a tough decision there's only one thing that outweighs every other concern, one thing that will make you give up on everything you thought you knew—every instinct, every rational calculation."

Dwight answers, "Some sort of virus?"

Jim shakes his head, responding, "Love."

I know. I know I know. It's horrible. In the context of the show, it also feels true. Ever fantastical in his thinking, Dwight imagines a zombie virus taking over, making us irrational. But no: Jim means that love expands what's possible for us, making us reject the confines of the rational. Hurston underlines Janie's love for Tea Cake; as long as she remembers him, he's alive. I wanted HIV from a lover, a connection between us, silent in me, to show how love and sex connected us. These stories are as viral as any other; viruses do live in us for good (a CMV like HIV), they can physically connect us to the dead, they can tie our molecules, our DNA, to our lovers. Why not these stories? Why not love?

In his book of essays, writer Alexander Chee wonders, "How many times have I thought the world would end?" In late capitalism, the world is always ending, both because of the desire—need—for never-ending growth, no matter the ecological costs or lost lives, and because crises drive consumption.

This is not the only way. Don't believe me? Look at viruses.

You can't see viruses from far away. We have to look more closely. Viruses are so different from one another. We have to tell one story at a time, to tell more than one story, to see that horror and death are not the only—or even the most likely—outcome.

Look at lambda. One virus, two options, two narratives, two stories, two metaphors. Destruction or symbiosis. Neither of these is the metaphor of *the* virus. Both are true stories of *a* virus. The difference between a definite and indefinite article. I don't think it's possible to imagine viruses without metaphors, small things they are, invisible to the eye. We need metaphors to understand complex things, and—I hope you agree by now—there are few living things as complicated as those nearly-living viruses. But we get to choose which metaphors we use, and choosing them wisely, carefully, with consideration of biology, of course, but also of our social relations, the systems in which we live, is (like lambda) a matter of life or death. Lysis or lysogeny. Death or symbiosis.

My friend Sarah died a viral death, but the way I think about her now is viral too. Like Tea Cake, her memory won't go until mine does. And her other friends, and her siblings, and her parents. With the sun for a shawl, I walk around New York City. I wear the pink scarf that her parents gave to me from her apartment in DC, bright pink even on the city's grayest days. She's living in me quietly, helping me along.

You're reading this, and so, in my mind, Sarah and her crooked smile live in you too.

Our work is not to get rid of viruses, or we would, by definition, fail.

Our work is to live alongside viruses and to protect as many human lives as we can. This depends, in part, on what viral stories we tell, what viral metaphors we use.

A virus killed my friend. I miss her every day.

I live alongside viruses every day, missing her. Her memory will make me smile; her memory will make me cry. It will make me angry, forever, at influenza, the virus that took her away, but that anger won't get her, or us, a second chance.

As individual humans and the collective we together form, death or symbiosis are our only options. The planet cannot continue to sustain our abuse. Will we eat it alive, use up its resources, and leave it an unsuitable host for further human reproduction? What will this earn us? Continued wealth for a small number of human animals is all. Human reproduction is not driving global warming; wealth production is. Human wealth will be lytic, killing our host planet and us with it. Lysogeny may still be an option. Symbiosis. We could understand, like one of lambda's stories, that treating the host well is treating us well. The earth's well-being is our own well-being. Lambda has its choice made for it by molecules and circumstances and luck. We have our molecules and circumstances, but we can make more than luck.

We must choose it, actively and every day, a lysogenic viral story, a living with and caring for the earth because it means caring for ourselves.

A virus is not an enemy; if it is, we will only lose. Viruses aren't the problem, they're a fact of the world. We are the problem when we refuse to protect one another's lives as the most precious things we have.

You are precious to me.

We might well prefer a world without viruses, their everyday annoyances, the fever or runny nose, the cold sores, the never-ending possibility of pandemic. We won't get a day without them. I miss Sarah every day. Viruses aren't going anywhere. We get to choose what we become. For my part, I live to be lysogenic. Won't you join me here?

4

On Private Writing

MY COVID-19 NOTEBOOK

Cast of Characters

LAILA: Tall, thin, beautiful; born in Cuba to a Black Cuban father and a Jewish mom from Chicago. Trained as a dancer through college, then did a PhD on paintings about poems and poems about paintings. Works in the art world.

NGOFEEN: Very tall, long forehead, and eyes that let you know what he's thinking. Got a law degree, worked in insurance litigation, had enough of his soul sucked, gave it all up to become a podcaster. Parents both from the Congo, born in the Midwest, has lived in New York for four years.

ANDREI: Broad chested, big shouldered, close-cropped hair, a beard of the same length; white and from Europe, born and raised in Poland before leaving for high school and college in England before moving to New York to do his PhD in the genetics of sensory neuroscience. Lived then in San Francisco, Germany, and Los Angeles before moving back to New York. Works helping academics patent outgrowths of their research for use in medicine.

DEVON: My bf, bae, partner, person. Short, with a huge smile. Hair cropped close on the sides, but grown out on top. Broad shoulders, skinny waist, skinny legs. He walks fast. Born and raised in Queens, New York; both his parents are Black Americans with family roots in the South. College in Vermont, grad school in California, works in social science research (designing polls and doing data analysis) for a marketing firm.

ME: Not tall, not short, dirty blond and dotted with freckles, especially on the face and shoulders. Raised in rural Washington State but from two Midwestern parents, of Irish and Norwegian descent. College in the Midwest, grad school in NYC, never left. Limp-wristed scientist; homosexual writer.

Monday, March 2, 2020 (Journal Entry)

At 6:15 p.m., sitting in my office cubicle—a fluorescently lit corporate-looking box that blocks off the view from the small room's large windows overlooking Washington Square Park—I texted Laila: "Wine o'clock?"

"Wine o'clock was 10 minutes ago."

"I'll be there in 15, wine in tow."

When I rang the bell, Laila's 9-pound mutt and my God-dog, Shiloh, greeted me outside her apartment door. I often stay at Laila's place—a tiny but lovely one bedroom in Chelsea—when she leaves town. From her place, I can walk to work at NYU. Shiloh loves to bury herself under the covers when you sleep, and sometimes, in her dreams, she starts biting your toes.

"So how are you?" Laila asked. I could only laugh.

"Are you going?" she asked. She knew I had a writing conference scheduled next week in Texas, one I go to every year. I'd paid for the plane ticket out of my own pocket, the Airbnb too.

"I don't know," I said. Standing just inside her apartment, I opened the wine bottle and she poured two glasses.

"Well?" she asked.

We talked about risk: I'm not the one really at risk, but people will be traveling from Seattle, where this virus is spreading, and from New York, where we think it's spreading, too. The risk is a conference center with 14,000 people from around the globe. It's a risk of spreading the virus more than the fear of contracting it myself.

"If it were me, what would you tell me?"

"Not to go," I said.

"I think you have your answer."

She talked about how, as a scientist, "a literal virologist," I have the ability to explain this moment to people. She said that this ability comes with a moral obligation to do the right thing myself, to show people what the right thing is, even when it's hard, even when there is little or no risk to me, even when I'd rather not. Shiloh was a circle on Laila's lap, I sat in the corner of her sectional, looking out the window.

"You're right," I said. "I'll write about it tonight when I get home."

I don't know what this journal will become. I don't know what's coming our way. I think, but I can't say for certain that it will be bad, be deadly, change our lives for a time, for a long time, for a year, for more.

When I got home, I texted Laila, "Made it!" and I texted my boyfriend Devon, "Just home crom chez Laila," and then "*from."

Here's what I think. What I know my other virologist friends think. The virus is here, in New York, today. Right now. In our communities. On the train. In offices. We—all the scientists I know—know this. And the city's doing nothing.

Well. Today I decided to do something. I won't be flying to Texas; online, I'll explain why. It's a little something, but I did it, even if that something wasn't much, wasn't nearly enough.

....

ESSAY ON PRIVATE WRITING PART 1 OF 6: COVID-19 DIARIES

By April we were already tired of quarantine diaries. They were in *The New York Times* and *The Atlantic*; they were in *The New York Review of Books* and *Slate* and covered in *The New Yorker*. Speaking of the Gray Lady, the paper published an article titled "Why You Should Start a Coronavirus Diary" (April 13, 2020) followed by a *Time* essay with a nearly identical title ("Why We Should All Be Keeping Coronavirus Journals," April 21, 2020).

I started mine March 2.

A journal or diary can help us process these emotionally dense times, to make sense of the senselessness of death, of loss, of isolation, of boredom, even as so much is happening at once. And journals and diaries will be useful documents for archivists and historians to come. It's plague journals (and fictionalized versions thereof) that give us a day-to-day imagining of the bubonic plague years in Europe (Daniel Defoe's *Journal of a Plague Year*) and the Spanish flu (countless historians' accounts).

But because this was 2020, the world moves at the pace of broadband. These journals became another news cycle. The Internet lets us not just write but publish in real time, no need for the two-year process of bookmaking, no need for years of archival research making meaning from primary documents. We sat at home anxious with nothing else to do than read about other peoples' anxiety; we were making (too much) meaning in real time.

Famous writers like Eula Biss and Nick Laird wrote about their lives. Stanley Tucci described his extended family—his young children and his older children, his wife and a friend of one of his kids, too. Tucci stole away moments to write "from my studio at the back of our garden in London." I picture him with tired eyes behind moon-shaped glasses, an early evening cocktail at his hand as he types—in my mind—on a burnt orange Olympic electric typewriter, the color matching a pocket square in his tweed jacket.

Healthcare workers—anonymous and named—described their horrors, their fears. The ill, including some writers like Leslie Jameson, detailed symptoms arriving, waning, racing back. The fatigues and fevers and shallow breaths, yes. They wrote to us from the vanguard of the virus in our country, and their testimonies were words hidden behind words: stay home, stay distanced, you don't want to walk this particular path.

Most of us who wrote in our diaries wrote without the urgency of a pressing fever reminding us of a deadly virus replicating in our cells. All of our daily rhythms became potentially deadly or impossible as decreed by law or necessity: get up in the morning, coffee shop on the way to the train on the way to the office, lunch with coworkers, dinner with a friend, a bar because you just find you have too much to talk about and dinner didn't let it all out. None of this was allowed, not anymore.

And so, to fill the space: the page. A new daily routine to replace the old, writing to replace—it would seem—life.

Inevitably, in 2020, writers wrote against this movement. In a book review on the first of the countless COVID-19 books to come, Lily Meyer wrote in *The Atlantic*, "No one has had time to truly refine their ideas about personal life in a state of widespread isolation and existential dread, and literature, even when political, is a fundamentally personal realm." Ummmm, can't believe I have to say this, but literature is always political and personal both.

It's impossible, Meyer is trying to say, to write well this close to a crisis; we simply haven't had the time for the remove that literature requires. Keep a journal. But for God's sake, keep it to yourself.

A journal, ostensibly, is for the writer alone; it's for our memory, or to process what we've just been through. And yet, journals, notebooks, and diaries—from famous creatives, yes, but also from regular folks—have been published as books for as long as books have been published.

Private writing, as I want to define it, moves across genre, from poetry to nonfiction to fictional mimicry. It includes writing that is seemingly for oneself, shared with the world. Journals, diaries, letters, notes from therapy, dream journals, notebooks. So many novels are built of letters or journals, giving a reader the sense of being inside the mind of a narrator, of seeing their thoughts, including the ones they might not ever share in dialogue.

But writing isn't just thinking on a page; it's also looking. Journals, letters, diaries, notebooks, these volumes often look at things otherwise considered mundane or unworthy of artistic attention. In an essay/interview on writing, Marguerite Duras describes watching, for minutes on end, the death of a fly on the wall of her home. This act of watching was writing itself: "Around us, everything is writing; that's what we must finally perceive. Everything is writing. The fly on the wall is writing." Even now, as I write on a Brooklyn balcony, I'm joined by a horsefly, a foot from my hand. Its eyes burnt red, it licks the table with its proboscis, sensing things I cannot see and therefore cannot describe. There is a tomato plant, now dying as fall marches in, and rosemary still growing strong; there is the sky, high and gray, and even a breeze. There is the fly, not dying, too much alive, but I don't shoo it away.

In Joan Didion's short essay "Why I Write," she says, "In many ways writing is the act of saying I, of imposing oneself on people. . . . It's an aggressive, even hostile act." Who gets to look at the world and write publicly about what they see is not random. It is impacted by educational level, gender, race, class, sexuality, geography, language, everything, all together, always. Many of the writers we read for their first impressions of our shared crisis were in the unique position of being able—by virtue of their profession and class—to get away. They wrote to us from country homes, they wrote of boredom, of having to be with their children in a way they typically are not, of having to care for themselves without the help. In *The Daily Beast*, Erin Zaleski wrote about this famous-writer COVID-19 journal backlash in the context of the French literary world: Can someone living in a 300-square-foot studio in the 17th arrondissement really care about the existential ennui of a wealthy writer in their six-bedroom country home?

With my own eyes, I can only see the facts of my own life; through private writing, I can visit the lives of others. Writing, as an art form, is unique in its ability to let us live inside another consciousness for not just the length of a film, but the time it takes to read a long essay or a book. In August, *Vanity Fair* published a collection of COVID-19 essays, including a journal by Kiese Laymon and a narrative about the death of her husband by Jesmyn Ward. These essays, exemplifying private writing, enlarged the number and type of COVID-19 narratives we, as a nation, could read.

Ward wrote, "The absence of my Beloved echoed in every room

of our house. Him folding me and the children in his arms on our monstrous fake-suede sofa. Him shredding chicken for enchiladas in the kitchen. . . . During the pandemic, I couldn't bring myself to leave the house, terrified I would find myself standing in the doorway of an ICU room, watching the doctors press their whole weight on the chest of my mother, my sisters, my children."

Laymon wrote, "Day Six: 9,400 Americans are dead from coronavirus, and Donald Trump will not say, 'I was wrong.' Mama is scared because the nurse we pay to take care of Grandmama will not wear her mask for fear that it could hurt my Grandmama's feelings. I am scared because Mama will not stop going to work. She sends me a eulogy she wants me to read if she dies. The eulogy confuses me. There is so much left out."

I needed to read Laymon and Ward. I needed to see the things Ward and Laymon were seeing.

Didion claims she became a writer because she failed academically: "In short I tried to think. I failed." For her, writing doesn't begin with thinking; it begins with looking closely. "In short my attention was always on the periphery, on what I could see and taste and touch."

"I write," she continues, "entirely to find out what I'm thinking, what I'm looking at, what I see and what it means."

In a crisis, in a pandemic, we all need to decipher what we're thinking, what we're looking at, what we see, and—most difficult to discern perhaps—what it all means. We all need to write. Writing cements the unbelievable in unbelievable times. Did that really just happen? Did we watch a deadly virus arrive in our country and pretend it would magically go away by April? If I write it down, I convince myself it did happen, even if it seems unbelievable. It was unbelievable. And it did happen.

....

Thursday, March 5, 2020 (Journal Entry)

Ngofeen is working from home even though his office is still open. I've insisted. He takes Remicade, which weakens his immune system, and so might be at risk for severe COVID-19 even though he's young and otherwise healthy. I tell him to tell his work he needs to stay home. The cases are high and still rising in NYC right now. We know they are.

Today we're working together from home. I live in Chinatown; he's in Flatbush. We're both home, together on FaceTime. We always have work dates, just sitting together writing, or me working on PowerPoint lectures while he cuts tape for a podcast. Now I'm at my writing desk in my little, light-filled bedroom, and he's at his kitchen table, and I have my writing up and he's cutting tape, and we both have our headphones in.

I can hear birds chirping outside his window through FaceTime. The birds sound almost like they're outside my window, but I live on a tree-free street, no songbirds to be found on my block, just street vendors singing prices and giving free orange slices below. His nature, the birds outside his window, a few miles away, through FaceTime, is mine. I'm anxious every time I go outside. I can't remember the last time I heard a bird sing. It is such a strange time to try to love and be beautiful toward one another.

Wednesday, March 11 (Journal Entry)

Ngofeen tells me that his office—where he works on an in-house podcast—is just swimming in COVID. Countless cases.

"Shit," I said.

"I'm glad you're not going in, that you stopped," I said.

"Yeah," he said.

"Who knows when I'll go back," he said.

And we both said nothing.

Saturday, March 21, 2020 (Journal Entry)

Does it matter that I write the dates in here? I'm writing in Google Docs, as we still have the Internet (for now). I'll be able to look back, 10 years from now, and see what days I wrote things. And I'm writing them as a journal, in order. What I write follows what I wrote before. This is never how I write essays or books.

My best friend from grad school just moved back to NYC from LA for a job. The city is finally shutting down tomorrow, Andrei got the keys to his new apartment today. For two weeks, he's been in an Airbnb looking for a permanent place. I biked to Andrei's Airbnb, to help him pack up. He'd sprained his wrist shopping for his new apartment, carrying bags, and so now he needed me to help with carrying bags. I didn't want to take the train—social distancing—and so I biked.

I decontaminated when I got to the Airbnb. I washed my hands, took off my shoes, redisinfected my fingers with ethanol.

"Can I have a hug?" he asked.

I offered an elbow. I figured he hadn't touched anyone in a week. Weeks? I knew he needed touch. But I was too scared to touch him.

Andrei today told me, after I helped him move, sitting on the floor in his new apartment, that this pandemic would change him forever.

"How?" I asked.

"Oh I just know it will. In a big way."

"Oh yeah?"

"Or maybe not," he added, laughing. "I just feel like the world will be different after, and me too."

I laughed. Don't we always feel the world will be different after, and me too. I think this each morning about the day waiting for me.

I sat as he finished unpacking. Our work done, we ate takeout from a Peruvian chicken place a half mile away. I'm happy to have Andrei back. I'm worried about seeing him in person just after he'd flown out here. It's been two weeks, but we don't know if that's enough time yet. But he asked for help, and that risk seems worth taking, to help a friend. Right? Our meal done, I biked back downtown, leaving him with a loving elbow tap. Remember, you can't touch your elbow to your nose, no matter how hard you try.

Thursday, April 2 (Journal Entry)

My phone rang between noon and 1 p.m. today. After work, I'm supposed to bike up to see Devon and stay for a few days at his place, since his roommate left town.

It was Devon calling me. It's not unusual to get a call from him in the middle of a day. Maybe he had tennis on and wanted to talk about a point. Maybe he saw a meme that was worth describing, and not just sending on, so he could hear me laugh. He liked talking on the phone, even if just for a couple of minutes, even if I was going to bike to his place in a few hours.

"Well, I'm done for the day," he said, and then there was silence. It was not even 1 p.m.

I was confused.

"That's great," I said. "I can come up now, we can hang a bit before my call!"

Again, silence.

I didn't know what I was doing, but I accidentally made him say it.

"I just got fired."

I owned the silence this time.

"Downsized, or whatever. Laid off. Because of COVID."

Fuck. I felt every inch of my flesh and it all felt bad, infected with an uncertain future, one I couldn't control.

"Let me pack my bag. I'll come up right now. Have you eaten?"

Silence, silence.

"Devon, have you eaten?"

"What?"

"Lunch. Have you had lunch?"

Silence, silence.

"Devon . . ."

"No . . . no."

"I'll bring you leftovers. Sit tight, see you soon."

It was cloudy and mild, something like spring, as I rode my bike as fast as I could up the Hudson River bike path. He lives some 10 miles from my house. The river looked gray to my left and the city buildings gray to my right and the green grass hadn't yet come up and so the path and park were muted colors themselves. They might as well have been gray.

Five floors of steps up to Devon's Harlem apartment carrying my hiking backpack full of food and wine, our needed supplies, and my bike. He opened the door. I was out of breath. He looked empty of himself, he didn't say anything, just turned around and walked back to the couch and sat down, pulling a blanket up and over himself.

Monday, April 6 (Journal Entry)

The rhythm of my weeks: Bike uptown to see bae Wed or Thurs, back home Saturday around noon to record a Food 4 Thot episode at 4. Saturday and Sunday prepping for the week. Teach Monday, therapy Tuesday, study session Tues night until 8 p.m., my God, teach Wednesday morning at 9:30 my good God, repeat.

The week starts now. I'm not ready. After work, I biked to Trader Joe's wine shop because lord. Lord lord lord. The line is now outside. We now have to take a cart, mandatory, "even if you're just getting one bottle."

On my bike ride home, I realized that it was exactly 7 p.m. when I heard the applause. From my own open window on my own block, I hadn't heard it yet. I was just a couple of blocks from home, and there it was, hooping and hollering too. Someone was pounding a pot, someone else pounding a smaller pot, higher pitched. It was odd because I couldn't see anyone. They were inside their apartments, and I didn't know which apartment the cheering was emanating from as I slowed my bike down the block.

I found myself choking up, and then I found myself wondering why. This gesture felt empty on social media. All the assholes who did brunch two weeks ago are now performing care for healthcare workers, clapping every night at 7 p.m. Honestly, fuck off. Biking down Mott Street, as the sun sets, two blocks from home, appreciative of the absolute absence of traffic, no cars to compete with, hearing clapping rise up suddenly from seemingly empty windows, why am I crying?

Writing now, I know I was crying because I was thinking of my friend Anthony—one of my oldest NYC friends, I met him my first year here. He's an ICU nurse. He texted me yesterday that he's drowning in the ICU.

"I just got through bawling on the metro north with my head down." He was heading back to the Bronx from NewYork-Presbyterian Hospital's overflow COVID-19 ICU. "This is now definitely one of the worst times in my life."

After the applause finally quieted, I was biking fast again downhill on Mott Street with a backpack full of wine and I was crying, for the applause, imagining it's all—all of it—for my friend, and I love him, and I don't want him to get sick.

I climbed the stairs to my apartment and wiped down each wine bottle with ethanol from the lab. I sat down again at my writing desk. My window open, the street is empty again. I can smell someone cooking on a floor below me. I find I am crying again, but now at least I know why.

....

ESSAY ON PRIVATE WRITING PART 2 OF 6: WRITING AGAINST THE PUBLIC/PRIVATE BINARY

Queerness, as defined by sexual desire and acts, has an odd, and old, relationship to visibility. Unlike some other identity groups, queerness invokes the possibility of a closet: it can be hidden identity. Women are usually (but not always) openly identifiable as women; race is often (but not always) apparent.

Each identity has its own way and history of malleability, and each is socially constructed. Gender, of course, has closets as they relate to performance and cis/trans and nonbinary identity. For queer and trans people, for many generations, and for many queers today, it's entirely possible (and maybe even likely) to live one's life in a quantum state: professionally in the closet for safety and survival, going to gay clubs at night; not out to family, building a queer family; performing femme at work and butch at the Cubby Hole or butch at work and femme at Pieces.

But on every gay first date I've ever had, we ask, "When did *you* come out?"

Queerness, though, is also a visible aesthetic. We do know it when we see it, don't we? Queer aesthetics, a rewiring of gender forms, assumptions, and expectations. Swish swish, limp-wristed men. Boy jeans, white tank leather jacket women. All the gender fuckery that is neither man nor woman but is decidedly queer, an aesthetic I've felt so much joy in embracing as I've grown up: heels and sports bras and T-shirt dresses and Doc Martens too. In other words, fem men and butch women and those who resist any binary categorization of gender are perhaps our most outwardly visible community members because we perform in every way we were taught—culturally—not to. And of course, this has nothing to do with our actual queerness, with how we identify or with our desires or sexual acts.

To openly identify as queer—inversely—is to invite the public into our bedrooms, into our desires and sex acts. As a visibly queer man, my identity says: I desire men. My identity doesn't lie; I put on fag drag on purpose. Like gender, desire is rendered personal only if it fits within what society deems acceptable. Judith Butler draws connections between laws against homosexuality and laws against miscegenation; both are meant to uphold a particularly racialized and gendered ideal of the nuclear family, and when those assumptions are troubled, both invite the state into the private sphere, the world of desire and sex, into the bedroom.

In her essay defining the term, Adrienne Rich begins with the statement that through "compulsory heterosexuality . . . lesbian experience is perceived on a scale ranging from deviant to abhorrent, or simply rendered invisible." In her readings of texts on gender and sexuality in 1980, she sees terribly little queer possibility. Lesbian desire—anything outside of a woman's need to have a husband and care for children—is not worth serious discussion.

It's worth noting that our legal right to gay sex stems from an interrogation of this question of privacy and a threesome gone wrong. In the case *Lawrence versus Texas*, the Supreme Court struck down that state's sodomy laws because the state, in the end, didn't have the right to know what kind of sex we're having. In the case, one man, jealous of the private flirtations of the other two, quite drunk, left to get a soda and called the cops, himself, to report "a Black male going crazy with a gun" inside Lawrence's apartment. One of the two men inside the apartment was Black; needless to say there was no gun. The cops decided to arrest both men anyway after bursting in looking for a "Black male going crazy with a gun" and catching the two men unexpectedly (to the cops) having sex.

Lawrence versus Texas ended with the legal right to gay privacy, rendering anti-sodomy laws unconstitutional. The state has no right to know who you have sex with. With gay privacy, our sex acts cannot be made illegal.

The year was 2003. The right to privacy (through which the right to butt stuff is born) for queer people is now 19 years old (it couldn't buy itself a drink at Stonewall). The right to privacy for queer people was borne out of an arrest that was based on the deadly ways of Whiteness, militarizing the police against a Black man against whom a white man was unjustifiably angry. Leave it to the cis white gays! The cops burst into the apartment—in Texas—guns first and caught the two men fucking. It could have ended so differently.

Queer people have since been left with a series of choices between assimilation into the American mainstream and continuing to insist on being sexual and sexualized in the public sphere. Legal rights have certainly been won by the former strategy, including the right to marry; many others—myself included—have troubled the water of gay assimilation by living outside of the expectations of the nuclear family: nonmonogamy, queer family, refusal to relocate to the suburbs and couple up and have 2.3 babies and a dog.

There is—as ever—a risk in visibly rejecting institutional assimilation, including by insisting on continuing to render the fact of queer sex (as opposed to queer love, which is love, is love, is love, is love) publicly.

Even before *Lawrence versus Texas*, queer people chose the risk of hypervisibility. In his first book, *Disidentification*, José Estaban Muñoz considered queer Brown representation on early reality TV, specifically Pedro Zamora on MTV's third season of *The Real World* in 1994. Reality TV, like private writing, promises us the impossible: a look inside real lives wherein people don't perform, don't know they're being watched.

Zamora was Cuban American, openly queer, openly HIV-positive, and an established activist when he went on the show.

"Zamora is willing to sacrifice his right to privacy because he understands that subjects like himself never have access to full privacy," wrote Muñoz. *Lawrence versus Texas* didn't yet exist; HIV criminalization was rampant. One didn't have a right to one's own body. Muñoz argues that his performance on *The Real World* continued Zamora's "life's work in HIV/AIDS pedagogy, queer education, and human-rights activism." Indeed, because viewers got to know Zamora, got to look inside his life, his love and marriage with a man, Sean, his relationship to HIV and his community, viewers, both gay and straight, had their realm of possible relations expanded. I was one such viewer, watching reruns once we finally got cable in the late 1990s in rural Washington State. I must have been about 16.

Zamora lived on camera in *Real World* San Francisco in 1994, moving into the house in February and moving out in June. Shortly after moving out, his health deteriorated. By September, he was admitted to St. Vincent's hospital in New York's Greenwich Village; by November, he died of progressive multifocal leukoencephalopathy, a viral infection of the brain caused by human polyomavirus 2. Ninety percent of the human population is infected with this virus; only when the immune system is nearly completely ablated does it cause any symptoms. The treatment for the virus is simply to get the white blood cell count up, and the immune system can start talking to the virus again, and the pathogenic part of the infection disappears, even as the virus itself stays for the duration of a life.

The last episode of his season of *The Real World* would air just before Zamora died.

Muñoz is careful to note that Zamora's use of corporate for-profit media, in this case MTV, to continue his work as an AIDS and queer rights educator is a complicated one; corporate media's interests are profit, not education. Yet, Zamora was able to engage with the mass audience of corporate media without altering his own reason for inviting the public into his private life.

Corporate media, since 1994, turned inverting the public/private binary into a massive capitalist project. In many of the franchises of reality TV, we see the (self-identified) 1 percent, but behind the closed doors they, by their nature, refuse to open to anyone outside of their social milieu. The housewives series, the Kardashians: these shows might be entertainment, and they certainly make us question what is private and what is public in the life of the famous and/or wealthy; they make us question what is real and false in reality TV and whether anyone can show the truth of their own life without scripting from a producer or making themselves more sympathetic (or villainous) to conform to expected storylines. I can't imagine anyone would call the Kardashians a radical project of public education, unless they mean only as a vehicle to stoke class resentment to the point of constructing a guillotine.

As queers become integrated into American life, Muñoz argues, "queers and other minoritarian subjects continue to be pushed into the private sphere." While Times Square used to be full of sex shops and cruising men, these have been replaced with "more corporate representation, such as Disney stores and Starbucks franchises." The "erotic economy of public sex" was stripped from the geography of the city, even as one's right to be gay in private expanded. Some Faustian bargain.

The public/private binary, Muñoz reminds us, "bolsters the dominant public." At the same time, not all exposure is radical. Capitalism's strength is turning radical gesture (like erasing the troubled notion of a private life) into something that makes money and changes nothing. Most exposure in capitalist media will bolster the dominant; care has to be taken to trade one's privacy not for capital but for radical possibility. Artists and activists who disidentify with the public/private binary can "perform activist politics." Rendering queerness public, in all of its relations, including sex, "challenges dominant protocols that relegate queerness, and other minoritarian histories and philosophies of the self, to a forced exile in the private sphere."

So many queer people fought and died to explode the nuclear family, not be incorporated into it. In his memoirs, *Close to the Knives*, David Wojnarowicz wrote of his terrifying upbringing within a typical American nuclear family. "I had a father," he wrote, "who brutalized his first and second wives with physical violence." Even as a child, he preferred the wild forest to "the Universe of the Neatly Clipped Lawn."

Rendering this private terror public through his work was essential for overcoming that trauma. Wojnarowicz writes: "[And my father] looks less monumental because I speak of him and bring the fear-charged memories of him outside of my head and make them public. . . . Words can strip the power from a memory or an event. Words can cut the ropes of an experience. Breaking silence about an experience can break the chains of the code of silence. Describing the once indescribable can dismantle the power of taboo."

Audre Lorde, in her *Cancer Journals*, wrote, "My silences had not protected me. Your silence will not protect you." Wojnarowicz reminds us that it may be the act of speaking that begins our walk away from suffering.

I write my queerness on my body as much as I can. I hid it in my youth, and resented that it wouldn't disappear. I write my life down on the page, all of it, the mundane and the erotic, the horrifying and the hilarious. Breaking the public/private binary in writing isn't in and of itself radical or revolutionary. It depends on what we have to show and why. Reifying the public/private binary by rejecting writing that's too messy, by claiming that writing is too personal, too sexual, too pornographic, "too much," this will always serve the status quo.

....

Tuesday, April 7, 2020 (Journal Entry)

In a faculty meeting on Zoom. One professor, as we waited for the meeting to start, said, "Listen up ladies—there's a touch-up feature to mute your wrinkles on Zoom." I open my video options and turn on the touch-up feature. I haven't been sleeping well. I've been sleeping, but it's not restful. I dream of infections and data, curves unflattened. I want to mute my wrinkles, too.

Friday, April 10, 2020 (Good Friday, Journal Entry)

The sirens. Day and night. They don't stop.

Sunday, April 19, 2020 (Journal Entry)

I'm texting with Ngofeen, who hasn't seen a person in over a month. He's stopped cooking for himself and doesn't get delivery and so won't really eat for a day or two, at which point his body melts down. His body is melting down now as he texts me, desperate for something to change.

Ngofeen: It's really easy to cook for others.
Ngofeen: It's really hard often to cook for myself.
Joseph Osmundson: You have to treat your body how you'd treat me or another beloved.
Joseph Osmundson: Laila told me that when I was going through my breakup with Wesley.
Joseph Osmundson: Because I felt the same, it was hard to care for myself in the way I could other people.

I look out the window. How will we get through this?

Ngofeen: I don't really know how to do that.
Joseph Osmundson: You have to trick your brain into it.
Ngofeen: I'm not sure I should ask the doc to
Ngofeen: Up
Ngofeen: The dose or something

He's talking about his Prozac.

Joseph Osmundson: You can ask, I donno what dose you're at.
Joseph Osmundson: But also cheri this is not a pathological thing.
Joseph Osmundson: This is you responding normally to BATSHIT CRAZY CIRCUMSTANCES.

My roommate is making dinner, and anyway, I'm not that hungry yet, and it's a warm day, and so I go for a walk. I don't have a place to walk to, so my feet just start walking. My feet keep walking and I find that I'm on my way to work. I haven't done this in more than a month, walked this path. I don't know why I'm walking here. Ah, I think to myself, I can go to Washington Square Park. My office window looks down on it. But, it's something like a green space, and I want to see how empty it can get. The park, year round, about the size of a city block, has a city-that-never-sleeps vibe, live music competing with itself from all corners and every place in between, a jazz band, a concert piano, a Bluetooth speaker. How empty is it now, with no students and the neighborhood on lockdown?

I walk up Mott from Chinatown into Soho. All the aluminum awnings are pulled down over shops. The only people I see are walking dogs, wearing masks. The city has a quiet that this city never has, especially not here, in Chinatown, SoHo, and the heart of Greenwich Village.

I feel good, I guess, maybe happy, walking this familiar walk. I've placed my feet here so many times. But, turning toward the park near the end of my walk, walking under the building where I normally work, I'm caught by an intense longing for the way life was, the normal way things were, my rhythms, my places, my geography of NYC before that geography consisted of two apartments: mine and my boyfriend's and the shortest, fastest, least populated line between them. In the park, I sat for 10 minutes on an empty bench before I heard a man, who'd been walking his dog, start coughing behind me and sit on the bench. With that, I shook my head, fearful now, and turned on my heel to return to the safety of home, my bedroom, tiny as it is, where own breath is the only breath. If there's a cough, it would be my own, no danger to me.

Tuesday, April 21, 2020 (Journal Entry)

Devon and I are moving in. We both need to get a place away from our roommates in this time when the home is all we have. Moving in a pandemic is hectic, of course, but NYC's real estate business has been deemed essential, both of our leases are up June 1, and we're ready to share a home. I think. We think.

Devon is . . . I don't know, to be honest. Last week, I got to his house on Wednesday night and he hadn't showered since the weekend. He stopped going outside. The blanket on the couch smelled of stale body. Devon was never like that. The next day, though, as I worked from his roommate's empty room, he stuck his head playfully into the door, hiding the rest of his body behind the wooden frame, until I saw him.

"Dance break?" he whispered, and I couldn't help but shut my laptop, get out my phone, put on Madonna, and dance. Madonna is one of his favorites; at 15, he took the subway to Madison Square Garden for her Confessions tour, alone. Devon has never needed friends to go someplace to dance.

He's on the phone now, listening to a busy signal. Calling unemployment has been a full-time job, looking for work has been a full-time job, and still he feels useless, like his days have no purpose.

"I'm going to be OK, right?" he asks me. I say yes. I have no idea.

"I've got some money saved," he says. A few months. We'd been planning to move for months, but now, could we? "I can't pay more than my rent now," he said. For a one-bedroom we might be able to make it work. We were hoping for a two bed, especially given work-from-home. Now? Not a chance.

We'll be all on top of each other. But at least we'll be together. And I'm making sure I can afford any place we pick all on my own for a while, maybe a year, just in case he's not fine.

I'm on StreetEasy in my home office, a little narrow nook delineated by a bookshelf on one side and a wall on the other, exactly enough room to fit my desk. I'm remotely touring apartments downtown and in BK, the sky darkens and begins to spit and the clouds flash and sing. The wind blows water in my window and I accept it on my skin, hiding my computer away instead of shutting the window up tight, closing out the outside world. The wind closes my bedroom door for me, the noise startling. I wish it were a ghost to keep me company. I want to take the outside world in, and this breeze is the safest way to do it, these liquid droplets carrying water, no virus, kiss me, mouth open.

There's a tornado warning, I see on FB, and I know from time in the Midwest that I should close my window and retreat to the inside of the apartment, a place with no windows at all. I take my risks, here, window open, wind whipped, staring at the sky even as it falls, which we used to call rain, before.

And then, infuriatingly, clouds gone in a second, wind dead, blue sky, the sun.

Wednesday, April 29, 2020 (Journal Entry)

I fall asleep on my side of his double bed and wake to Devon holding me for warmth, his arm on my chest, his leg around my leg, and like this—wrapped in him—sleep comes back to me again.

Sunday, May 3, 2020 (Journal Entry)

I'm writing this on my phone from a field in the north end of Central Park. After helping Andrei move into his new apartment, seeing him in person then, I decide to see him in person again.

I biked to Andrei's house, the first friend I'm adding to my quarantine pod. My pod: me and Devon, Andrei and Ngofeen—my friends living alone, who I think need the company—the touch—the most. Laila has her bf; we can FaceTime just fine.

I dropped my bike at Andrei's house and we packed our bags for a picnic in the park. On the way up, I stopped by Bagel Works, on 1st and 60th, a block from where we both lived together as grad students. One of the best bagels in the world. Him: sesame flagel with half veggie chicken salad half chicken dill salad. Me: everything bagel toasted with the same mix of chicken salads, his invention from 2007, but I add red onion and tomato. We walk from his place on Lex to the Central Park baseball fields, where there's plenty of space to safely distance.

The sun on my skin. The breeze too. Sitting 4 feet from a friend. Writing on my phone. Andrei pours me more wine. My first year in New York, Andrei brought me to Central Park on my birthday, a snowy February afternoon, for mimosas as we walked. Now, the bagel the wine the grass. My allergies coming up just like any other spring. My eyes itch and water, just like any other spring, Andrei rests his head on my leg and sleeps as I write this in my phone. It could be any other spring. The sparkling rosé with strawberries tastes fresh, tastes good. The sun burns my winter-white flesh, nothing else, nothing here, I'm lying on a blanket with my best friend, his head on my leg as I write this, just like any other spring.

Friday, May 15, 2020 (Journal Entry)

Got keys today. Brooklyn. Prospect Park two blocks away. Parkside Q train. Balcony massive. Roof amazing. Can see the Empire State Building, downtown Manhattan from our roof. The living room is a little tight, but I think we'll manage to fit bar stools, foldable table, bookshelf, couch, coffee table, chairs, and office, all in a big long row. I measured everything out and put painting tape on the ground to give me a sense of the layout. Devon's uptown, I Face-Time him once I get inside. Picture the big open room empty now, well lit with sunlight through the window, blue tape mapping out where the couch will go, the table, coffee table, and chairs. I think it will work. I think.

Saw Ngofeen today, first time in two months. He moved to this neighborhood from Flatbush six weeks ago. He wanted to say hi now that I'm almost moving here too. I mean I feel like I see him all the time. We've FaceTimed more than anyone. Ngofeen lives just two blocks away from our new place. We kept 6 feet apart. He walked me to the park, Prospect, only a stone's throw away. I knew it wasn't all the way safe to stand next to him, even outside. It was good to see him. I hadn't forgotten that he's flesh. And that I am too.

Friday, May 22, 2020 (Journal Entry)

It was a TV mount that finally broke me. I don't buy myself new things when I don't need to. I've had the same TV for three NYC apartments. It's a fucking TV. I don't need a bigger one, it works just fine. I've mounted it myself in the two apartments I lived in before: The Village and Chinatown. Now, in BK, I'm confident in my skills. The TV is the same but I splurged on a new mount; the old screws had stripped. I tap the wall and listen. I first think the walls here are concrete after a good amount of tapping. But then the stud finder finds 1.5 inch studs exactly 16 inches apart and I think YES! This will be so easy, then! Finally a New York City wall with wood studs! I know how to do this.

I did not know how to do this. Things seemed off from the start. Last night drilling the pilot holes, I hit resistance (yes!) past the drywall (yes!) but then am through that to emptiness (oh no). But the new long and large mounting screws don't have Phillips heads on them (probably smart, a socket wrench head prevents stripping), and it calls for a 1/2-inch socket and my tool kit maxes out at 7/16ths. Of course. So I go to bed, holes in the wall.

I walked to the little hardware store here one block away and buy the exact drill bit I need for under two bucks. Things seem . . . good. I get home and I drill. I hold the drill in my right hand and push, stabilizing it with my left, my arm straining to generate force, and then relaxing, and then pushing into the wall again. Devon watches with a skeptical look on his face. He is not a fan of the holes in the wall. “Don’t worry boo, I got this.” Things seem . . . bad. It’s not wood that I hit, of course, I finally notice, but what? Concrete? The mount has a setup for that too, it needs larger holes, and so I drill one, try to pound in the wall anchor with a hammer as described and only end up breaking it. I borrowed the hammer from Ngofeen, who’s sitting and helping, now, in a chair. He carried the hammer over to me.

“I remember this happened at my old apartment,” he said. “But that’s where my memory stops. I don’t remember how we fixed it.”

I laugh. But now there are three big holes and one very big hole in the perfect walls of this perfect apartment. I’d said that I could do this, that I knew how, and clearly I couldn’t, and I didn’t. There’s a flush in my body. There’s a feeling in my gut, a sickness not from a virus or bug but from anger to and at myself.

All this death and all this misery, all this time cooped up, alone. All this fear, this diligence, this washing of every vegetable and package in isopropanol as soon as it comes inside the apartment, and waiting for two days for all the things we can wait for: dried pasta, blue corn chips, avocados, onions can sit out two days, nothing but dead virus left after that.

“You’re just going to have to wait for help,” Ngofeen says, as I stare up at three large holes and one very large hole in a once perfect wall. He’s texting his contractor friend and the guy he found on TaskRabbit who does light construction stuff in and around his home.

I lie down on the couch. I cannot move. All this death and misery, and I am gutted and ruined by three large holes in the wall, one very large hole in the wall, an unmounted TV, concrete shavings on the floor, and my utter and absolute inability to do anything about it. All I can do is lie here and try not to cry.

"Fuck," I say.

"Yeah," says Ngofeen.

"You're just going to have to wait for help."

"I know how to do this," I say, unsure of myself. No, I'm finally sure that I do not.

....

ESSAY ON PRIVATE WRITING PART 3 OF 6: PRIVATE WRITING AS A RESPONSE TO COLLECTIVE PAIN

Global catastrophe with its individual suffering seems to beget private writing. Marie Howe wrote about her brother dying of HIV in *What the Living Do*. Paul Monette wrote about his partner dying of HIV in *Borrowed Time* and *Love Alone*. Marguerite Duras wrote about her husband being taken away to a concentration camp in *War: A Memoir* and *The Wartime Journals*. Charlotte Delbo wrote about her own time in a concentration camp in *Auschwitz and After*. David Wojnaworviz wrote *Close to the Knives* about his own death of AIDS, which he knew was coming.

We write so much about catastrophe because we're trying to make meaning of death at a scale we have a hard time comprehending. One death can send us spiraling, but a million?

In the TV show *BoJack Horseman*, Henry Winkler explains a particularly odd caper that BoJack undertakes after the death of his friend. "You ascribed a mystery to Herb's death to give it meaning. But there is no meaning in death. That's why it's so terrifying . . . There is no shame in dying for nothing. That's why most people die." Of course this is true; of course we cannot accept it. Of course we turn to words and stories when the world seems otherwise impossible to explain.

War: A Memoir—a work of autofiction by Duras—is an odd

translation of the French language title, *La Douleur*, literally translated to "The Pain," which I might translate to "Pain" or "On Pain." Its twin manuscript is the notebooks Duras used to compose the novel. The novel came out in 1985, decades after the end of the war; the notebooks were published after her death, in 2006.

In her *Wartime Notebooks*, we see raw writing, ostensibly written in real time, a diary of waiting after the end of the war for her husband to return or not to return:

"I'm tired. The only thing that does me good is leaning my head against the gas stove, or the windowpane. I cannot carry my head around anymore. My arms and legs are heavy, but not as heavy as my head."

"I can't go on. I tell myself: something's got to happen—it's just not possible . . . I should describe this waiting by talking about myself in the third person. Compared to this waiting, I no longer exist."

We also see, in her notebooks, the mundanity of war: "La Rue de la Gaieté on Sunday afternoon. People are coming down the street with the sunshine at their backs. All shops open: direct communication between the crowd and what's in the shops. The girls' heavy legs. Boys' jackets nipped-in at the waist."

This mix of horror, of fatigue, and of seeming—what?—normalcy? Shopping on a Sunday afternoon, it could be any Sunday afternoon, and you can almost feel the sun? How to face this all, and all at once? It's best simply to write it all down.

Duras faced the immensity of death and loss as the war was ending. It became apparent how many million civilians had been killed in the camps, and how writing can, will, must, cannot confront it. "Much has been written about death," she writes in her *Wartime Notebooks*. "It's the chief inspiration for the artistic endeavor. The face of death discovered in Germany, on the scale of eleven million human beings, confounds art. Everything comes up against this crime and struggles against this giant dimension no cross could bear. . . . I think of all our poets, of all the poets in the world, who are now waiting for peace so they may sing of this crime."

This all feels so familiar, now. We're always exhausted, and bored, too. We're faced with an immense horror that necessitates art, and yet art feels impossible in the face of such massive collective death.

In his memoir of the last years with his partner, who died of AIDS, Paul Monette describes sending poems back and forth in the mail to a friend. "I was writing with a very blunt instrument, but groping toward leaving a record—'to say we have been here.'" Monette wrote his memoir with the closeness and linear time line of a diary and included journal notes as well; we're reading the inside lives of two men fighting to live, but knowing in many ways that their lives are unlikely to continue.

From his journal: "Lying next to Rog in the guest bedroom. We went out for supper to Cock and Bull and took a walk down Cory St. I never thought I'd write in here again, I never thought I'd do anything again, but I record with gratitude and a sense of calm that we stepped out for a plate of prime rib."

He wrote, "I note in my journal that we ended the evening by floating the notion of being grateful for a good day."

Health, in those with AIDS, waxes and wanes, and with every good week, Monette wrote how "here we were, agreeing again to the lie of normalcy and holding out for veal chops."

Monette's work reminds us that relative wealth won't protect us from suffering. He and his partner—a lawyer and film writer—have all the markings of gay affluence: a house in the hills, two sports cars, veal chops on a weeknight. And despite their Whiteness and despite their wealth, together they died of AIDS.

The lie of normalcy. How else can one survive a pandemic? Who is safe during one? The lie of normalcy. How else can one survive a death that takes years to arrive?

"Loss teaches you very fast what cannot go without saying," wrote Monette. And so we write it all down.

When the bad days came, Monette found himself comforted by his ability to care for his beloved, his little friend. His partner Rog suffered a common symptom of AIDS at that time, a detached retina and eye damage due to a herpes infection. After the surgery, Monette wrote, "Finally, something to do. And when you do this part you come to see that there's something nearly sacred—a word I can't get the God out of, I know—about being a wound dresser. To be that intimate with flesh and blood, so close to the body's ache to heal, you learn how little to take for granted, defying death in the bargain. You're an instrument, and your engine is concentration. There's not a lot of room for ego when you're swabbing the open wound of an eye."

Caring for his partner was a way of begging for more time with him. Something to do, some way to help: "It turns out a home-cooked meal offers a double dose of magic. At the same time you're making somebody strong again—*eat, eat*—you are providing an anchor and a forum for the everyday."

Oh the calm I get from cutting garlic and adding it to onions already in a pan, a dinner for me and Devon, an end point to work for the day.

With all this pain, these daily moments of joy and also horror, absolute horror, Monette found it was often the mundane annoyances of life that sent him spinning into despair. In *Borrowed Time*, it's a possessed sports car: "On January 30 the Jaguar—possessed by a devil, clearly—locked in gear again in the parking lot. I proceeded to go bananas, an instant replay of four months earlier. Roger screamed at me to call Triple A and leave him out of it."

Reading this, I see myself so clearly. Like for Duras and Monette, writing feels impossible, and yet it feels impossible not to write. Like Duras, I wonder how art can say anything about the deaths of millions of people, as COVID-19 and HIV have both killed millions by now—as of today 33 million dead of HIV, 2.5 million dead in the first year of COVID-19. Like Monette, I find writing a compulsion, a way to say that we were indeed here, that this did indeed happen. To write down the ways we cared for each other—a swabbed eye, a home-cooked meal—and the ways we had to pretend life—for just one second– was normal.

The weight of a million dead in months I cannot not feel. I look up and am surprised to remember the sky is still blue. I don't know how to weigh the mass of my own personal horror this year, or how to compare it to others. I know that facing this horror, I've needed to keep a journal, I've needed to write it all down. In her notebook, Duras wrote, "Why remember the movie theater all of a sudden? Urgent to write it all down."

Urgent to write it all down.

Kumeyaay poet Tommy Pico, in his book *Nature Poem*, wrote "my great grandparents had almost no contact with white ppl like the shutter of a poem is the only place where I can illusion myself some authority // Everyone remembers the weather when discovering a body // I think it's perfectly natural to look skyward."

Paul Monette wrote, "Evenings at the brink of summer are yellow gold across the city's western face, as the sun narrows toward the ocean, eye to eye with the white buildings of the coastal basin."

Duras wrote, "People are coming down the street with the sunshine at their backs."

Primo Levi, chemist and Holocaust survivor, wrote about watching his dog playing in the mountains, even as fascism swept through Europe: "He aroused a new communion with the earth and sky, into which flowed my need for freedom, the plenitude of my strength, and a hunger to understand the things he had pushed me toward."

Delbo wrote, of her time in a German concentration camp entirely devoid of color, "A break in the clouds. Is it afternoon? We have lost all notion of time. The sky appears. Very blue. A forgotten blue."

Tommy Pico wrote, "The air is clear, and all across Instagram—peeps are posting pics of the sunset."

A million dead in nine months. I'm doing everything I can. Except for everything I can, all I can do is write. I look up. I write. All I can see is blue. I close my eyes to mute its beauty.

....

Sunday, May 24, 2020 (Journal Entry)

What's the opposite of crying? Today the full quaranpod assembled for the first time. Ngofeen and Andrei came over to our apartment at the same time. None of us had had haircuts since February, but we were finally all together. Ikea is on backorder, so my books are still piled in boxes, but we moved them into the pantry to have them—at least—out of the way. Devon hates having guests unless the house is in perfect shape, but even he is happy for company given . . . well, given everything. It's almost Devon's birthday. I grilled, the grill itself a hand-me-down from Ngofeen. He'd brought it from Pittsburgh when he moved to the city, but never had a place to use it. Now, I have a balcony. He told me to take it. I told him to come on by any time he wanted to grill.

Yesterday, I bought veggies (onion and red pepper for a soy-based sauce, asparagus for a white wine–based sauce), pork tenderloin (a simple dry rub of kosher salt and rough-chopped rosemary), and burgers (I used 2 lbs 85/15 ground beef, an egg yolk, jussssssst a pinch of bread crumbs, ample salt and fresh cracked pepper, and 2 slices of bacon just cooked through, with the rendered fat).

The TV is on the wall. Ngofeen and I hired a handyperson to mount it, a professional would know how, and it would be well worth the 60 bucks.

Andrei arrived first, washed hands, took off shoes, put on house shoes, got out a cake for Devon, washed the cake box, washed hands. He asked if he could sit—he took the train—and then, answering himself, said no.

"I got on at 96th"—the Q—"the first stop. I could feel the seat wet with sterilizer. But still."

He picked up our can of Lysol.

"Maybe if I spray myself."

"Girl."

"What?"

"Just stand outside. Sunlight kills it."

This is not science fiction.

He stood outside and first sprayed the entire back of himself in Lysol, sticking out his ass.

"Hello Brooklyn, I'm here!" he sang off the balcony, faking a Broadway voice in his slight Polish–English–New York accent.

"I'm putting on a showwwwwww!" his note high and then low, lingering, vibrato-filled.

I laughed. Devon laughed. Andrei came inside and yes he did sit down. I fired up the grill for the first time. First on: the veggies and pork. They're good finger food once the pork is cooked and sliced nice and thin. In the end, the doneness would be something I was proud of, just a touch of pink at the center.

Ngofeen showed up with homemade biscotti, and I tried my best to get everyone's burger doneness right on. I did OK, not perfect. I made gimlets for the guests, and I loaded the dishwasher, and I was outside cooking and I could hear the laughter inside and feel the sunshine on my left arm and feel the grill's heat on my knuckles, searing.

Is this normal? Is this the new normal? I don't know, I don't know. But today, with three people I love in the house, cooking for them—one of my favorite things to do—my body released the tension in my back, my shoulders, the tightness at my temples, the tightness of my jaw. After we ate, Ngofeen fell asleep on the couch. Devon went in to nap on the bed.

"Are you kind to yourself?"

I looked up.

Andrei looked over at me. He had that serious look on his face. He wanted to talk. I put my phone down.

"I think I am."

He looked at me sideways.

This is what it was like before, this January, earlier this same year, when I was in Los Angeles in his apartment on his couch sipping his champagne. He picked me up after a long flight and now there we were, here we are.

"I know things are good with Devon," he said today. "But how are you doing with you? Are *you* OK?" He kind of knew what to ask, even if the question seemed obvious. Isn't that what friends do?

"Are you being kind to yourself?"

"Well, I have . . . it's just I have such high expectations for myself. So I'm hard on myself. But not . . . not unkind."

"But isn't that unkind, in a way, too?"

"Shut up Andrei," I said, and he laughed.

"Look. Work has been helping me survive. I feel good when I'm working. Or when I'm writing. I don't know if it helps anyone but me. When I take a break, my anxiety takes over."

"I know, like on Friday," he said—two days ago—"I wanted to take a half day at work. But I got into it and I was enjoying it and . . . and I just kept working. Who was I going out to see? What was I going to do?"

I looked down at my empty glass of rosé.

I looked back up at him. Him. A human person, a fleshy body. He always knew the right question to ask. His body, here, and the breath we share so dangerous, and yet he's in my home, and yet he's loving me, and yet, and yet, and yet, he is entirely worth the risk he poses, and I couldn't be happier to have taken this risk. I look down at my empty glass of rosé and my jaw unclenches, the first time I realized that it was clenched to begin with.

Wednesday, June 3, 2020 (Journal Entry)

Noon, Wednesday, humid and overcast and here I sit on the patio drinking iced coffee and trying not to start sweating.

What can I say?

Daily protests following the murder of George Floyd in Minneapolis. Monday night an 11 p.m. curfew. Yesterday, word of curfew at 8 p.m.

"That's not even dinner time in New York," Devon said. I found nothing to say.

"Pray for me," Ngofeen said. His months of isolation due to quarantine and then the emotional weight of videos of police violence and then of protest, and now the added weight of a citywide curfew for almost a week. We'd been planning to cook together tonight. We decided, now that we're neighbors and now that we're in a pod, to cook together once a week, on Thursday.

"We'll have to have dinner early, like 6:30, so I can be home by 8."

"Amore, I hate this," he said.

I found nothing to say. I hurried to his house to cook together, to be together, at least.

Late night, a thunderstorm. Devon and I turned off the TV and sat out on the patio, shielded from the rain by the patio above. We counted seconds between the lightning and the thunder, never counting fewer than eight, never seeing the rain become a downpour.

"I love a good storm," Devon said. "I haven't seen lightning in ages. Not just the flash," he said, "but the big lightning, you know, across the sky."

He sipped his wine. It was 1 a.m., Twitter told us the protesters made it home, that there wasn't as much violence as Monday night, that there was no looting.

"Who gives a fuck about looting," I said.

Quiet. A flash of light, and, 10 seconds later, a low rumble across the New York City sky.

Monday, June 22, 2020 (Journal Entry)

If anyone reads this, it will be months or years from now. News happens fast these days. I've been calling these times dense, and they are. Emotionally dense, with loss and struggle and even sometimes joy. Scientifically dense, with papers and pre-prints out every day that need reading and some analysis. Like a firehose in your face, we've all said more than once. News dense too. Just the last few days: Trump's Tulsa rally; Juneteenth; Trump and Barr fire U.S. attorney in New York who was investigating . . . Trump. Black Lives Matter protests continue. Thirty articles about the 4,000 percent increase in fireworks in cities like NYC, Boston, LA, everywhere, it seems. Fireworks from 8 p.m. until 2 a.m., Macy's quality fireworks, somehow everywhere and all at once.

Remember all that?

What were you feeling, then?

What am I feeling, now?

Today, I listened to Tommy read his essay on *Catapult*, "Is this shortness of breath because of the Rona, or because of anxiety from living through A GLOBAL PANDEMIC?"

Yesterday, I made fried green tomatoes with a garlic and basil dipping sauce, a play on a recipe from *Bon Appetit*. The tomatoes were on sale: heirlooms from Whole Foods. I fried two that I kept in the fridge, green, and blended two overripe for the sauce. The fried ones were hot and tart and salty, crispy and gooey, both. The fresh sauce hummed with garlic throughout, cool. I cooked for friends—my quaranpod—the only two people I've seen face to face since Laila, March 2.

Immokalee, Florida, is a tiny community in the south central region of the state. Today, Doctors without Borders is running a mobile COVID-19 clinic, testing and treating the farmworkers, the essential workers, who are getting COVID-19 by the score, by the hundred. They're essential workers, so they're working. They have no PPE.

They're growing tomatoes.

Immokalee has 24,000 residents, 74.1 percent of them Hispanic/Latinx. They reported 175 new cases last week. The county has reported more than 1,200 cases, 15 percent of the documented COVID-19 diagnoses in the state. Florida—across the board—is underreporting COVID-19 cases to save political face, not lives. The nearest hospital is 40 miles away from Immokalee in Naples, Florida.

America is Immokalee.

Outside of New York, NJ, and Connecticut, COVID-19 never waned. Now, a second wave after the first wave never stopped coming. Now, today, states across the country running at 70 percent or more capacity in their ICUs, full again of critically ill patients struggling to breathe, patients on oxygen or ventilators to fill their lungs up and keep them alive.

Yakima, Washington—5 hours from where I grew up and an agricultural center for apples—is out of ICU beds, they are airlifting patients to Seattle. They grow apples in Yakima, and hops for brewing. Seventy percent of the apples in America come from Yakima, 20 percent of the hops in the whole world come from near my home.

Immokalee means "your home" in the Mikasuki language of the Miccosukee and Seminole peoples. Your home. Yakima is named for the Indigenous people who fought a war to stay on their land, their home, but lost and were removed to the Yakima Indian Reservation, 2,186 square miles of rangeland, home to 32,000 people and 15,000 wild horses that my family looked for as we came down I-82 south of town, my head craning out of the window for the telltale cloud of dust.

I am America, eating tomatoes on unceded Lenape land, tomatoes I've wiped with isopropanol before frying in a pan to remove any virus that might have been coughed, sneezed, or breathed onto them in the store.

What am I feeling now? What was I feeling taking a bite, yesterday, of a tart, hot, creamy, salty fried green tomato? Ngofeen was over. Andrei was over. Devon was home. I made comfort food for us all. I watched Ngofeen take a bite of his tomato, witnessed his look go from curiosity to pleasure. "Now try it with the sauce," I said, like I'd have said a year ago. What am I feeling? What would joy be like to feel again? No. No. I'd die if I could feel anything at all.

....

ESSAY ON PRIVATE WRITING PART 4 OF 6: WRITING PRIVATE ILLNESS

It's not just public catastrophe that drives us to write. A private catastrophe, one in the body, can do the same.

In 1976, Susan Sontag sat dying or not dying, and she wrote. Sontag had cancer. She wrote about cancer in *Illness as Metaphor*. She wrote as she was herself ill. The big C. She wrote, "Today, in the popular imagination, cancer equals death." She wrote, "As long as a disease is treated as an evil, invincible predator, not just a disease, most people with cancer will indeed be demoralized by learning what disease they have."

Writing seems a reasonable reaction to the possibility of physical disintegration, to the threat of the annihilation of any version of herself not requiring a faith in the hereafter. Writing is forever. This is a human quality, one that becomes stronger as the end feels inevitable. Perhaps this is the writer's reflexive response to trauma in the world around us or to trauma within our bodies. We don't understand, and we are afraid, and we feel alone, and so we seek to explicate, if only to ourselves.

She did not write a journal. She did not write a diary. Publicly, she wrote an essay.

She never claimed her own cancer; she didn't admit in this book that she was ill, facing her end. She could not write herself well. She could not write away her illness. She wrote away the place her illness gave her in society, a double illness in her mind. She could only cure the word *cancer* of its myths, which were as deadly as the disease itself. She wrote, "Fatal illness has always been viewed as a test of moral character." She didn't think she was being tested; she was just sick.

In 1979, Audre Lorde sat dying or not dying of cancer, and she wrote. She wrote for herself, private words. Her work announced itself as private in its very title, *The Cancer Journals*, when she shared those journals publicly. The stigma of disease, the pain of recovery, the fear, these forces can choke us into silence. She wrote, "I was going to die, if not sooner then later, whether or not I had ever spoken myself. My silences had not protected me. Your silence will not protect you."

Lorde wrote against and into her pain, against and into her death, against and into her body. Lorde saw other women around her fighting the same disease, but longed for another Black body, a feminist body, a lesbian body, to pull her along. And so she bared her body, her illness, the possibility of her death, publicly. In 1979, breast cancer looked like it did to Sontag in 1976 and to my own grandmother in 1973. Lorde was Black, which made any illness look different, which made and still makes breast cancer more deadly, not because of the biology of race but because of racism.

I know stories of my grandmother's 1970s cancer from my mother watching *her* mother's pain, from the matriarchal lineage of memory in my family. I know my grandmother's almost death, even as her children were just beginning their adult lives. Major surgery, perhaps followed by chemo and/or radiation, or perhaps not.

Lorde was coerced into wearing a puff of lamb's wool in her bra immediately following her surgery so that her missing breast wouldn't be apparent to others. My grandmother had a silicone implant put in so that she would look like a "normal person" (her own words) in clothes. Lorde wrote, "I looked strange and uneven and peculiar to myself, but somehow, ever so much more myself, and therefore so much more acceptable, than I looked with that thing stuck inside my clothes. For even the most skilled prosthesis in the world could not undo that reality, or feel the way my breast had felt, and either I would love my body one-breasted now, or remain forever alien to myself." When my grandmother's implant burst inside her, the pain was worse than the surgery, worse than the cancer itself. Her daughters caught glimpses of her body rarely, in a mirror. I don't know if her sons saw her body at all. I know this because her daughters, both, told me.

My fear of cancer, of the ultimate and forever annihilation, of the moment at which all feeling ceases, made me miss one fundamental truth of this particular disease. My research on cancer, how I turned it into a biological problem to be taken apart, to be solved, hid this from me. Cancer hurts. It just fucking hurts. Our bodies are not our own to control. After my grandmother died, I read Audre Lorde on her own cancer, her own almost death. She said what my grandma would not, what my grandma let pass in silence or only told to her daughters. Lorde sat dying or not dying from cancer and she wrote, "There were fixed pains, and moveable pains, deep pains and surface pains, strong pains and weak pains. There were stabs and throbs and burns, gripes and tickles and itches." My grandmother told her daughters, "It is a pain of 10 out of 10 to lift my neck from the pillow. It is like having a 100 pound weight on me." My grandmother asked me, "Why am I still alive?"

Lorde's private writing showed me things I didn't know about my own family. This is the power of private writing, of publishing journals, even while we're living: this level of intimacy and vulnerability models a new way of being in relation not just to writing, but to one another. Our relationship with the text can change how we want to treat, and be treated by, other people, including our family.

Sontag explained her choice not to include private writing, or even the details of her own identity as a cancer patient, later in another essay, this time on the metaphors of HIV: "Twelve years ago, when I became a cancer patient, what particularly enraged me and distracted me from my own terror and despair at my doctors' gloomy prognosis was seeing how much the very reputation of this illness added to the suffering of those who have it."

"The metaphors and myths," she writes, "I was convinced, kill."

She didn't talk about her own cancer in her book. "I didn't think it would be useful," she said, "to tell yet one more story in the first person of how someone learned that she or he had cancer, wept, struggled, was comforted, suffered, took courage . . . though mine was also that story. . . . A narrative, it seemed to me, would be less useful than an idea."

But Sontag's ideas came from her body. For me, as a reader, feeling *with* Lorde, and seeing the ideas that came from that feeling, changed my mind and body. As a reader, the journal showed me more. Sontag claimed that telling her cancer story would be narrative, common, something we've already seen, a story we already know. But I see it differently. It's not *just* narrative; it's embodied feeling. And it's worth it to feel and think at once.

In her book *Funeral Diva*, Pamela Sneed writes of attending funerals of Black gay poets lost to HIV in the early 1990s, almost too many to count. But it wasn't just Black gay men dying of AIDS, it was women, too, like "Pat Parker / The pioneering Black lesbian poet who hailed from San Francisco / [who] like Audre Lorde had died prematurely from cancer." Parker died in 1989. Audre Lorde died in 1992, at only 58, of breast cancer. Queer Black lives lost from HIV, queer Black lives lost from cancer. A body is a body is a body; White Supremacy produces Black death. How many lives and words were robbed from us not by cancer or by HIV but by homophobia, by racism?

Sontag and Lorde: I cannot read one of these two books without immediately reaching for the other. They are siblings in my mind, twins whose differences seem so stark because the underlying circumstances are—by definition—so similar. Two women had cancer and wrote it down. One woman wrote an essay, never naming the fact of her cancer. The other wrote a journal, naming it over and over.

José Muñoz considered Pedro Zamora's decision to live his private life publicly on TV, writing, "subjects like himself never have access to full privacy." Lorde, a self-described Black, feminist, lesbian, understood that publicly writing the full extent of her private illness was a radical breaking of the public/private binary, a binary that for her—as a queer Black person—had always been a lie. For queer people, as Muñoz writes, privacy is a recent and incomplete right. For Black people, the history of America is one without the possibility of a private life: what right does property have to privacy? Resistance to this dehumanization is a history of languages invented to remain whole as people and a white American public that either violently reacted to or appropriated these languages into the mainstream. Lorde is, in her book, speaking for herself and on her own terms, sharing her life and world in exactly the ways she wants. She might not have had me as a reader in her mind as she wrote, but I am so immensely grateful that she shared her private life with me in ways that Sontag didn't seem able to.

Sontag's book looks exclusively outward, at the world of literature, of ideas. "A narrative, it seemed to me, would be less useful than an idea." Private writing requires, in my thinking, an ethic and aesthetic of looking inside, of laying the body or life of the writer bare, of looking closely at the self, and claiming that self on the page.

If you write it down, it never needs to make its way into public writing; if you don't write it down, it will never have the option to. The rest is a question of revision, of editing, of choosing what to share. The vastness of life requires editing. You can't live alongside me; you—the reader—have your own life to live. So what moments from my life birthed ideas? If I want to share boredom with my reader, what bored moments of my life should I focus on and write through? What moments distilled a feeling that meant something to me, or that I learned from?

Our ideas come from somewhere. Lorde's famous notion that our silences will not protect us, quoted so often and almost always without naming *The Cancer Journals* as its source, comes from the experience of being a Black lesbian breast cancer patient.

My ideas come from reading Sontag, reading Lorde, and living.

....

Saturday, July 18, 2020 (Journal Entry)

Optimum commercial: "It's never been more important to stay up to speed."

Sunday, July 19, 2020 (Journal Entry)

I almost forgot to water my plants on the balcony. It was nearing 100 today and the pepper's leaves were wilting. Guilt. I can barely keep my plant babies alive.

It feels immensely important to harvest the fruit of these plants, the rosemary and sage, the hot peppers and patio tomatoes.

But why? There's no way I could live off patio tomatoes, or any numbers of patios full of tomatoes. I can't sustain myself this way. My parents, when I was younger, had to grow much of their own food, otherwise we wouldn't have been able to eat, but they had enough land fertile enough for more tomato plants than I could count. One year, rot. Another year, pests.

All other tomatoes are outside of my control. They will either arrive at the grocery store or not, where I'll be able to afford them or not. These tomatoes, these ones, I can control, I can foster and care for and eat in return.

An acre of patio tomatoes won't save me now. I can't grow enough to fill my gullet. I rely on the outside world to live.

For 5 minutes, before bed, I can water my plants and pretend that this is all I need. I picture the salad I'll make with these fruit: basil and tomatoes and burrata, purchased too pricey at the store. I taste it on my tongue, now before bed, and imagine that this—here—is all I need.

Wednesday, July 29, 2020 (Journal Entry)

I woke to COVID dreams. First, a party. Not a gay club or anything like that, but a party maybe for a family graduation or reunion. Family, neighbors, friends from childhood. We were in a home that wasn't my family home but was familiar. I could get back to my parents' house by walking. Halfway through the party, I realized what I had done:

I'd hugged my mom and sister and a family friend. I looked up and around: People eating and drinking and socializing with no masks for the first time in months. Oh no. What have I done? Devon will be so mad at me. *I* am so mad at me. Why did I do this? How could I forget?

I don't wake. The dream goes on. I can't leave the party and I can't leave the dream. I want to scream and shake everyone, but I don't want to disturb the peace, and so I stay and smile and laugh along all while screaming inside and knowing, just knowing, that this is how I finally catch the virus.

COVID dreams: The tomatoes I've been growing since May finally ripen, the whole plant of them, a wealth of tomatoes, my favorite fruit. In one day, they rot, spoiling on the vine.

COVID dreams. Sex with a person I find online. I know the COVID risk. I'm not on PrEP. I go over to their house and I knock on the door and I am on my knees and then I'm in their bed and they're in me. I shake with risk more than pleasure. I shouldn't be doing this, I know. I shouldn't be doing this, but I am. It's the risk I feel, only the risk, and the room falls away, the bed comes out from under me, but the man is still in me, something like floating, something like falling, and that's when I knew I caught the virus, but I wasn't sure which virus I'd caught.

I wake.

Wednesday, August 19, 2020 (Journal Entry)

Devon is in a job interview. His second this week. The only two he's had since April. I've been scared to write about it, but our household has been . . . I don't even know the words. Devon hasn't had work for five months now, other than the work that is looking and applying for jobs. It's been horrible for him, every day just looking and looking. No people to work with, nothing actionable to get done, no reason to leave the house.

And he hasn't been himself. He's been having a hard time getting motivated to get out of bed, to exercise, to fuck, to cuddle. He's depressed. And it hurts me to see him so unlike himself.

"I just need to get a job," he told me in June.

"Do you think I'll have a job by September?" he asked me then.

"Do you think I'll have a job by the new year?" he asked me last month over dinner.

"You can only do what you can do. And I'm proud of you," I said in return. "Everything else is out of your control."

He's in the living room at his desk, and I'm in the bedroom at my own. I can hear the person interviewing asking him questions, now, through the door. "I just want to hear a little bit of your story." I can't take the stress of listening so I put in earbuds and blast Chopin's nocturnes, which I've used since high school to calm myself down. I used to play them, now I just close my eyes and remember what it felt like to be able to make those sounds grow from my fingertips.

This week and last, preparing materials for these interviews has made him more himself. He had to get a deck done for one of the interviews, and he worked on it for hours. He was proud of the final product, and happy when—within just hours of it being sent—he was offered a phone interview with the department chair.

He's been waiting to hear about next steps from that interview, yesterday, as he prepared for this one, today.

Do I dare put a bottle of champagne in the fridge? Not for today, obviously. I don't think any of these jobs are ready for a dotted line today. But just to have waiting? If I do, will that jinx his chances? I want this so badly for him. I chill no wine, sparkling or otherwise.

The interview is over, and I walk into the living room. I sit down at the kitchen island to write. As I sit here writing, he's put his food in the microwave, but hasn't turned it on. It looks like he's waiting for his food to turn and heat, just standing there, but he isn't.

"Babe are you OK? Is the microwave on? I don't hear it."

He pushes the +30 seconds button three times quickly then.

"Sorry, I was just thinking about my answers."

I don't know what to say.

"Let it go, eat! Do you want a glass of wine?" It's 2 p.m.

"I do," but he doesn't get one. He sits on the couch with his now warmed food, leftovers from last night's salmon in spinach, Congolese style like Ngofeen taught me. I return to Chopin. I turn back to writing. He puts on *Housewives*. We both disappear into a world a little easier than this one.

Monday, August 24, 2020 (Journal Entry)

I'm waking up, dozing still, and not in my own bed. I'm in a beach cottage, yellow walls the color of outside sunlight, the A/C too loud to hear the waves.

Devon is awake already.

"Why are you up, babe. It's early." It's 9 a.m., but we're on vacation.

"I heard about the job," Devon tells me. This is supposed to be vacation. I don't want this part of the world in my world yet, but here it is. The world won't leave us alone.

Vacation in a pandemic. I've been working, writing, publishing, teaching, for so many months. Usually my summer would grant me a two-week vacation, some time to recharge. I traveled, each summer, to Los Angeles to visit friends and plant my feet in the Pacific. Andrei lived there for the last five years. Now he lives here, in New York, and I'm too scared to get on an airplane. In LA, I hiked. I sipped rosé outside on a friend's patio under fuchsia-flowering vines. I cooked and cooked and cooked, feeding all my friends who woke up with work to go to. I wrote.

Devon and I haven't left the city since March, which—for me—means that my travels in January were my last travels of the year. We moved in May, driving from one end of the city to the other, from Harlem to Chinatown to Brooklyn. Our new home has a balcony and a roof; we live three blocks from Prospect Park. I'm tan, my childhood blond peeking out in my hair. Of all my New York summers, this is the summer I've spent the most time outside. This is the longest time in years I haven't been outside of New York, not even for a weekend.

We've been two scared men in the little world of our making. The walls of our apartment haven't felt like they're closing in on us. But they aren't expanding either. The world is largely stasis: Devon looking for jobs, me trying to stick my finger in the dike of our government's response to COVID-19, and to write, and to prepare for a semester of in-person teaching that seems destined to arrive. A stasis of constant crisis.

Devon didn't get the job.

"They wanted someone with implementation experience," he said. I had no idea what that meant.

That night, holding hands, outside. The sky is punctured with white dots. Walking at night on the boardwalk or in the street, the stars look countless, or so they seem. I know, as a scientist, they can be counted, but I don't want to be a scientist. Not now. I want to look up and sip my wine and stare at the light coming from stars, old light. I know the universe is expanding, and these stars are getting farther away, even now.

I need a rupture, one I control. I need to check my email once a day, and leave it at that. I need to sleep late in a bed that isn't my own. I need to say "Fuck it! Why not?" before pouring a glass of wine at 4 p.m. Fuck it, it's 4 p.m., and I'm drinking a Pilsner.

"Am I going to be OK?" Devon asks me.

"Yes," I say. I'm shocked at the sound of my own voice. I believe myself, almost. "It was your first interview. It was just practice, it's about process, not outcome. I'm proud of you." I believe what I say.

"I have to keep reminding myself," he said.

Last week, back home, when I was listening to a clip of Trump on Twitter, he said, "Can you please turn that off. I can't even hear his voice." A pause, then. "That man ruined my life."

I turned it off. "That man is a trash heap," I said. "But your life is not ruined."

Now, the sun streams into the lofted bedroom in a yellow-walled beach house.

"Good thing I'm here to remind you. You're going to be more than OK."

I almost believe myself about him; I almost believe myself about all of us.

....

ESSAY ON PRIVATE WRITING PART 5 OF 6: AGAINST NAVEL-GAZING

Confessional writing, what I'm calling private writing when it extends into journals and diaries and other private documents, is an aesthetic, not a genre. Aesthetics are political, of course; an aesthetic without politics is fascism, as initially noted by Walter Benjamin in *The Work of Art in the Age of Mechanical Reproduction*. Susan Sontag was right: all aesthetics have politics, even ones that claim to have none, like camp.

In the case of confessional writing, as I noted before, we are allowing the public sphere access to the pleasures and horrors of private life, particularly the private lives of women, queer people, people of color, and other minoritarian subjectivities—minoritarian not in a democratic sense but only in relationship to the majoritarian culture that is cis hetero male and white. The nuclear family is a construct that both renders affairs of the family unit private and makes labor forces more "flexible." Economists say frictionless. Physicists know that friction is inevitable; friction shows you that you, and the world, both exist.

In the nuclear family, child and home care are done by the woman; the whole family can pick up and go wherever the jobs may be. Life's messiness is the friction: the kids like this school; we want to be close to family and friends; the woman, too, has a job. The two-body problem, needing work for not just one but both adults in a household, is friction. Pulling back the curtains to look closely at everything we're told not to speak of? Sex, child-rearing, the insides of depression and anxiety? This, on the page? This is, in its very bones, deeply political work.

By the end of the 1970s, Joan Didion was famous enough in the literary world to have a target on her back. A famous takedown by Barbara Grizzuti Harrison, "Only Disconnect," takes Didion to task for her self-regard and what is read, by Harrison, as cold remove. That Didion's work can be both obsessed with the self and also too sterile, too cold, too unsure for Harrison speaks, I think, to the intellectual emptiness of this particular line of critique. But read it anyway we must.

"No; in fact, her subject is always herself," Harrison writes. "In the 1960s, she says, 'no one at all seemed to have any memory or mooring.' But to what is she moored? What she is moored to, of course, is her angst. And her angst is not the still point of the turning world."

"Didion is the lyricist of the irrational," Harrison writes. "Some people find that charming. I do not."

Didion, by her own definition, is writing to find out what she thinks, and to share that discovery with her reader, even as she knows that insisting that her reader care is a massive ask. We need essayists indeed who write from a starting point of knowing; Didion is not that essayist, but her work dives into the place where anyone claiming to know with certainty is lying. We need writers like this, too. Ones willing to write about the world falling apart and include notes from their shrink confirming that they were falling apart, too.

Harrison critiques Didion on her politics, this being a solid foundation for criticism, finally. Didion's lingering conservatism colors her early work indeed, and her essay on the women's movement in *The White Album* is frustrating at best.

Harrison writes, "Many of Didion's observations about the self-serving 'children' of the 1960s are dead accurate; but that doesn't give her the right to fiddle while Watts burns." I agree, partly. But Didion's writing on the ways in which the progressive movement itself was fracturing is a necessary work, too. Harrison continues: "It's true that Didion occasionally ridicules the rich; it ought not to follow that this gives her the right to express contempt for the poor. I call that writing sentimental; I call that sensibility nasty."

There's a lot, in terms of class and race, to trouble in Didion's early work, yes. But to call her sentimental, to call her nasty? Can we imagine a man would be treated this way? Or someone writing more conventional, outward-looking journalism?

One final quote from Harrison: "I can't resist quoting something Gloria Steinem once called out to a journalist on her way to interview Didion: 'Ask her how come, if she spends all her time crying and swimming and struggling to open a car door, she finds the energy to write so much?' "

I want to slam a car door on my hand. One can write about one's fragility; the act of writing doesn't make one strong. Brave, perhaps. One can write about paralyzing dread, tears that overtake a body; the act of writing does not negate the lived experience of horror. One can write, horrified. One can write while crying.

Writer Pauline Kael, in the pages of *The New Yorker* in that same era, called Didion's *Play It as It Lays* a laughable book, hooting at the first and last sentences both, and many in between. She accuses Didion of "spiritual emptiness," adding that "what's missing from these books is the morally tough common sense that has been called the strongest tradition in English letters."

I want to slam a car door on my other hand. Questioning and queering the tradition of "English letters" is exactly the point. Being unsure is exactly the point.

Writer Elissa Washuta moved Didion's impulse to include a note from her psychiatrist to a new depth in her collection of linked essays *My Body Is a Book of Rules*, which contains an extended letter from her mental healthcare provider and a diary of psychiatric drugs taken. Other essays in Washuta's collection include blog posts and entire IM conversations, a collage of real things, digital ephemera from a life lived, right there on the page.

"I think I am going to keep a Dear Diary," she writes on December 22, 2006, at 12:25 p.m., "because I like to write things down. Private things. I can't tell anyone what's in my head."

These confessions and the inclusion of private writing—notes, diaries, IMs—create a kaleidoscopic image of bipolar disorder and/or trauma, the only way—I think—to understand what it's like to live alongside these painful internal realities for a long life. How can we occupy any other brain, not to mention understand trauma and its impacts? The workings of a brain are never linear.

The irony here in an earnest one: Washuta can barely write in a diary, her thoughts are so personal, so "private." And yet here they are, public, in a book. I don't know if it's brave, I don't know if it's self-indulgent; it may well be both. As a reader of Washuta, I know I, too, have private things I can barely write down. I have a psychiatrist and take Lexapro, something I've been taught—in a million ways I cannot name—to hide from the respectable world. I take Lexapro, 20 mg, because I've had insomnia since I was 10 and anxiety since before I can remember and sometimes it makes it hard for me to even be in this world.

Susan Sontag wrote that "to be a moral human being is to pay, be obliged to pay, certain kinds of attention." I argue that it's worth paying attention, if we can stand it, to the inside things, private things, hidden things, body parts. In *A Room of One's Own*, Virginia Woolf reminds us that "so long as you write what you wish to write, that is all that matters; and whether it matters for ages or only for hours, nobody can say."

Confessional writing always risks these accusations (Pornography! Melodrama! Nasty!) as if all of our lives didn't include sexual organs and what we do—or don't do—with them. As if life isn't melodramatic! Worse, perhaps, than accusations are eye rolls, the expression "navel-gazing" used as a reason to not engage in the first place.

Who am I to look? Who am I to write? All art is self-indulgent. We're requesting the time and attention of our readers and viewers; who are *we* to believe we are owed this right? Who am *I* to claim my life is so interesting that you'd want to read my journal?

Melissa Febos also writes about how things come easier to her in writing: Things that can be impossible to say in life, to an ex or a lover or a mother, suddenly become not just possible on the page, but urgent, necessary. This is work we do first for ourselves. Why share it with the world? "My darkness has become my work on this earth," Febos writes. Others have struggled with queerness, with consent, with addiction. "You think your pain and your heartbreak are unprecedented in the history of the world," James Baldwin wrote, "but then you read."

My life is not so interesting. All lives are mostly boredom, enough horror to render us wary of having children, grief and joy and drugs and sleep and dreams and nothing and nothing and boredom and nothing. It's worth it to write it down and then try to make it urgent and enchanting enough to trick people into reading.

....

Tuesday, September 8, 2020 (Journal Entry)

My alarm goes off today and I have somewhere to be. Not just somewhere metaphorically. I have a meeting. Not a Zoom meeting. This is a meeting on campus, the first since March, six months ago. I'm not just biking in to work to sit alone in my office.

I'm biking in to work to prepare my webcam for teaching from my lab tomorrow. I'm biking in to meet—face to face—with the TAs from the other lab I teach, to show them the lab equipment we'll be using, and get the webcam set up for them, too.

This is the world that's been set up for me. I had little choice in it. This class is in person—decided by the university—and I'm one of the managers of the TAs, and the TAs need to be trained, and so here we find ourselves, four bodies and four hearts and eight lungs and four mouths dangerous to one another.

We sit 6 feet apart. We sit in masks. I show them which buttons on the GoPro make it actually go. Natasha is one of the TAs, and we worked together in lab for two years before I took this job teaching. In a way, she's like family to me. Back then, we saw each other every day at work. Our desks were touching. We've been on a group chat this whole time, and for years, and so I know what her baby girl looks like.

"How's Gabby doing?" I ask. She shows me a picture on her phone, holding out her hand so our bodies can stay far apart.

I'm surprised; it feels good to be with people. It feels nice to be meeting. I'm cognizant, always, of how many feet there are between me and the other people in the room, and I imagine the air in the room slowly filling up with our exhaled breath, a potentially dangerous thing. I haven't been inside this long with anyone but Devon and Andrei and Ngofeen, and that came after negotiating a testing schedule, talking about what risks we'd take outside of our little group. We promised to check in, even, before hookups.

Back at work, I calculate the volume of air in the room and the number of minutes we've shared that air, but who cares, it feels good to talk to people without a screen between us.

What I'm scared of is what feels good. Plagues place our bodies at odds with our minds.

An hour later, I'm in a Zoom room with my therapist, Dr. Eric. I'm in my office, he's in his home. We see each other on screens.

"When do you feel completely safe?" he asks me.

"When I'm at home lying on the bed with Devon," I say. "At that moment, I feel completely safe."

"What about now," he asks, "in your office, alone? Do you feel completely safe?"

"Well no, because I had to come here, ride an elevator, meet with TAs . . ."

"But right now. Right now in this moment. How do you feel? What are your risks right now?"

"None," I admit. None.

"Just be mindful of those moments, and let yourself really feel them," he says. I try to scoff but I find tears in my eyes.

I tell him that even moments of pure joy and clarity have felt risky, even if that risk is invented. When I cook, as I run my knife through onions and garlic or fish or chicken, I often picture the knife cutting through the animal layers of my own flesh. My hand slips under the already moving blade, a deep cut running red before

I feel anything. I can see it in my eye and I have to pause. Cooking has been therapy for me these months, but even in this safe space, I feel the risk. I see the risk becoming reality. I know that in my long life so far, I have cut myself cooking, badly. Nothing is safe, not even this.

These bloody imaginings started in April. If I cut myself cooking, then, I could get COVID at the hospital. But I couldn't not cook: most restaurants were closed, even for takeout, then. Fuck. So I pulled my fingers back, made sure the cutting board was dry, and still I imagined a steel blade working open my epidermis, spilling out my blood. For once that blood was not my plague worry. My plague worry: the air I'd breathe as my flesh got sewed up.

It's nighttime now and I'm home again with Devon and I'm writing.

"What are your risks right now?" Dr Eric asked me.

"None," I say to myself now. "None, I'm just in my home and writing."

Maybe—I see now—this is why I'm writing so much. When it's just me and an empty page, I can conjure a virus without being susceptible to it. The virus becomes a thing *out there*, and my *in here* becomes safe.

Praise be to writing, something I can do alone. Something I can do alone, with the world, Amen. Blessed be this writing, I don't know who it will save, but it's already saved me. Sing to heaven of this writing, it's here that I feel completely safe. Consider all the worlds thy hands have made, this page is the world I can control. O my soul praise the banal joy of a meeting of human bodies, and let us leave it unscathed, amen. Tune my heart to sing. Four mouths and eight lungs, keep all of us safe, our mouths, our lungs, our hearts, oh hear. Oh words, let my lungs fill with air, let you be expelled and breathed in, through our dangerous air, and let us all wake tomorrow alive, but more than alive, alive, breathing deep, alive and, too, well.

Friday, September 25, 2020 (Journal Entry)

The most basic beginning of a journal entry: Sorry I haven't written in a while. I've been tired. I've been busy. I haven't felt like it. I haven't been up to it.

This is my lab notebook. This is what happened today. I'm writing it tonight. I'm writing it as it happened. As it's happening.

I'm not up to it now. But this lab notebook needs filling out; I need to remember.

Many times in my life I haven't been up to writing, and I've still written. Emails and texts, of course. But also essays and a book proposal.

Stuff happened today.

It's 1:22 p.m. and I'm watching someone I love become a doctor. She's defending her PhD. She and Natasha and I all worked in the same lab, together. I'm on Zoom on my computer and I have my camera on because I know how much easier it makes presenting when at least a few people in the Zoom room are looking back at you. My email is open but invisible. My texts are open on my computer. They're visible. I'll admit something I don't like about myself: Once every 5 or so minutes, I click over to Chrome, check my email, check FB notifications, check Twitter and what is trending there. Devon is on the roof doing yoga or a workout tape, I don't know which. It's a nice day, a cool fall day, full of sun, the type that makes you search for the light and the heat it brings. Blue dot text message.

Devon Wright: Good news! —— is working on offer documents! The HR woman is working on getting approvals from the UK!

I read it again. And again. I lose track of Zoom, the words coming and going without being heard.

I read it again.

Joseph Osmundson: !!!!!!!!!!!

I read it again.

Joseph Osmundson: NO FUCKING WAAAAPOY-JEOIUTNEOIUFNWIUFNWIM CRYINGGGG-GGGGGGGG Im buying a nice champs for tonight BABE!!! It's not just good news!!!!!!

I turn off my camera on Zoom, and I stand up, and I look up at the ceiling, and I imagine the sky, and I let out a shout, no words just a yelp, and then a second, one loud and long word, "Yes!" My fists are clenched.

I turn my camera back on and I'm back in the room where a good friend is becoming a doctor.

Five minutes later, Devon texts again.

Devon: I'm coming back down now.

When he opens the door, I turn my camera off. He doesn't want a hug, he's all sweaty, and so I make him give me a high five, and I am the type of happy where it's coming out of my feet. I'm dancing, little steps, saying "yes yes yes" and he's smiling and dancing too and we're holding hands and Max is barking because he doesn't quite understand what it means when Devon and I are playing like this just with each other but not with him.

And then I click back on the camera and I'm back in the Zoom room watching a friend become a doctor. How much magic is one day allowed to hold? And will all this magic hold? Devon hasn't gotten the paperwork yet, but they promise him it's coming next week.

I'm tired, but grateful. Grateful that my partner has filled up the work-sized hole in his life and budget, knock on wood until the contract comes through. Grateful for friends who are even now becoming doctors right before my eyes. If others before me can smile . . . I can laugh in rage and joy simultaneously.

Wednesday, September 30, 2020 (Journal Entry)

Tomorrow is October, can you believe it? COVID is spiking in NYC, but slowly. What will the cooler months bring?

Devon signed on the dotted line today. He's employed. Andrei called this morning. He's been back home in Poland a couple of weeks, trying to help his parents out. Things didn't go as planned. His mom had a health crisis. She had to go to the hospital. In the hospital, and then in her ward, COVID. Now his mom has COVID-19.

"I'm the one who made her come to the hospital," he tells me. She was negative when she arrived.

"She'd already be dead at home," I tell him the truth he already knows.

No easy joy this year, even knowing that other people and other people's parents are ill. But now, it's him, his parent. Andrei. My Andrei. Life is happening, and so death is too. He's back home in Poland. I can't even risk a hug, a risk I would—now—happily take.

....

ESSAY ON PRIVATE WRITING PART 6 OF 6: SAY WHAT YOU MEAN

I host a podcast where I talk about fisting. I write essays, ones that are posted online, about safe sex policies at sex parties where I make it clear that I go to sex parties. In my first book I talked about not being clean when my ex-boyfriend was eating me out.

All these things are about my life, and I claim that on my podcast, in my essays, and in my books.

In June, Andrei told me, "At least you're having sex," because I had a boyfriend in quarantine. I told him I wasn't, and he responded, "At least you have someone to touch you," and yes, that was true.

There's no sex in this diary because I haven't been having sex. It hasn't been a thread in the fabric of my daily life. Devon's been completely uninterested in sex; out of work, he's felt himself unworthy, he says, of that type of pleasure. I've been anxious, too, and only slightly more interested in fucking than he is. This has never happened to us, or to me, but it's a pandemic, and I'm trying to be easy on us.

Writing about not having sex is scarier for me than writing about sex itself, something I've gotten used to. Both break the public/private binary. Both feel dangerous, but the former feels embarrassing. I've lost it. My quaranbod is sagging. He doesn't find me cute anymore. I'm getting old. It'll never come back.

What kind of gay have I become? I've been in relationships before, longer ones than this, and we've only had sex *more* as we stayed together longer. So many partnered older folks I knew told me sex would wane as I grew older, or at least wax and wane in the context of long relationships. "*Not me!*" I thought every time, without saying it.

I'm reading T Fleischmann's *Time Is the Thing a Body Moves Through* (a reference, by the way, to the four dimensions of space-time) for fun. I'm on page 26, and there have been at least six orgies. I love Félix González-Torres too—one of the two subjects of Fleischmann's book, the other being sex and how sex and friendship are compatible for so many of us queers—but my celibate summer is making me feel like a bad fag, or even—God forbid—straight, as I sit reading this queer book with Devon napping, legs on my legs.

Am I even gay if I'm not having gay sex? Yes, I mean of course; but in my own estimation of myself and my sexuality, I'm not living up to the identity I used to make for myself. O–, growing older is tiresome, and it seems to never end.

I guess what I'm trying to say is that Foucault's notion of homosexuality as the possibility of friendship is what my life has looked like. In an interview late in his life, he said, "Perhaps it would be better to ask oneself, 'What relations, through homosexuality, can be established, invented, multiplied, and modulated?' The problem is not to discover in oneself the truth of one's sex, but, rather, to use one's sexuality henceforth to arrive at a multiplicity of relationships. And, no doubt, that's the real reason why homosexuality is not a form of desire but something desirable."

Less fisting, less fucking, more friendship. Or both. You can fist your friends; I have. I love fisting and fucking, I'll have them back some day, I'm sure. Hooking up with strangers or friends outside the pod feels dangerous. I don't know. But looking at these pages, I'm reading the tides of my life, and sex isn't a part of them, not now.

I can't go back and write in sex, even though I want to, because it wasn't a part of the truth of my days. If it were, in here it would already be written.

Looking at the tides of one's own life is immensely useful. Friendship bubbles up on these pages; I knew that, but I know it more now.

And this journal might look like it's mine, but it's only interesting because it is not. Here, in these pages, you see deep into some of my relationships, those friends in my pod.

If everything is writing, a dying fly is writing. We write with our actions. This document, this journal, is not mine alone. So I can't publish it alone either. But these pages contain my family. A life I have not lived alone. And so, their permission to make their private lives public is necessary for me. And we agreed to share these parts of ourselves. I can only thank them, my family, for living and writing with me.

Disrupting the private/public binary is a risk, but it's worth it. I grew up young and queer and in the closet but only because I didn't know that bisexual people even existed. I didn't know I was possible. I was afraid of sex because I knew nothing about it, or pleasure, or consent, or how to say yes, or how to say no. The inability to speak about these things is ruinous.

And so, learning from Lorde, I speak.

Even as it's still dangerous to write publicly about gay sex and relationships, about the insides of queer lives, those of us who do so are still mocked online for navel-gazing. It's much cooler to write from a distanced, ironic, sarcastic remove. In a 1993 essay on irony, consumerism, and American fiction, David Foster Wallace defined what was then the fashionable way to write: satire of American consumerism. But then satire and irony became tools used in advertisements and TV; how can writers critique ironic consumerism with irony, he asked?

His answer: we have to say what we mean once again.

"The next real literary 'rebels' in this country . . . might be the ones willing to risk the yawn, the rolled eyes, the cool smile, the nudged ribs, the parody of gifted ironists, the 'How banal.' Accusations of sentimentality, melodrama. Credulity. Willingness to be suckered by a world of lurkers and starers who fear gaze and ridicule above imprisonment without law."

Private writing takes up Wallace's task: It is single entendre. It's to Wallace's own shame that he didn't realize that many people—mostly women, mostly queer, mostly people of color—were writing earnestly in America all along. He wrote in 1993 from a world where he could read Sontag. Audre Lorde had just died and Marguerite Duras was still living. He wrote while Joan Didion was publishing essays at the *New York Review of Books* frequently. And he thought it wasn't *trendy* to write earnestly! Can you imagine? It's no surprise that so many of the women, people of color, and queer people were sneered at for writing that seemed too interested in *the self*.

There's nothing more banal than a diary. There's nothing more predictable than a COVID diary. The year 2020 reminds us that life is melodramatic, full of impossible coincidence and relentless boredom.

On June 22, I wrote, "I still can feel." I could then. I still can now. On October 5, 2020, Andrei's mom died of COVID.

Maybe, writing that, in the early summer of 2020, I knew COVID-19 would take Andrei's mom by fall. When Devon's grandma got COVID-19, we weren't surprised. She had so many preexisting conditions—her age and her dementia and her scarred lungs. Her home health aides had been coming and going for nine months. The vaccine was available, but she was homebound, and I couldn't seem to help get her an appointment at home. She died the following January, 2021, only minutes after Devon left her side, after we rented a car and I drove him through the falling snow to her hospice on Long Island. I stayed in the car. Only one person was allowed inside. We knew this pandemic would come for our family. And yes, here it was. Yes it did. Maybe, writing that in the early summer of 2020, I knew COVID-19 would take Devon's grandma before New York's snow had melted from the winter. Collective grief comes for us all. I knew it would come for us, too. I still feel, I still feel. I knew it was just a matter of time, of statistics, of case counts, of good and back luck.

On March 21, 2020, Andrei said to me that this pandemic would change his life. "Oh I just know it will. In a big way," he said. I know because I wrote it down. On April 18, 2021, at his birthday party, having just read all this, he said it did. How do I know? I know because I wrote that down, too. He gave me a big hug, like he always does. "My mom is gone," he said, "my life has a hole in it and it will forever. My life is changed."

My life is changed too, of course it is. Seeing that and saying that may be the first step toward accepting this, what our life is now.

"I didn't want it to be like this," he said. "I didn't think it would be like this." We both held champagne in our hands; some funeral this was turning out to be, some birthday. Bubbles rose, in my hand, out of the corner of my eye, bubbles rising in my eyes, and his eyes.

I can still feel. I'm so glad I remember. I'm so glad I wrote. Devon is family now. Andrei and Ngofeen are family. Their grief belongs to me, at least in part. I can report that I feel it. I feel their joy, too, and our joy when we are together builds and builds and builds until it becomes almost impossible. My body shimmers with gratitude. For better or for worse, my loves. We are a family now, pressure forged, and that's what family means. Until death do us part. But let it not come too soon.

5

On HIV and Truvada

An Intimacy Only I Had Earned

In late 2014, I received an email from one of New York City's largest sex parties for gay men. Usually, the email would have contained this: a time, an address, a dress code, the price. The party had long been condoms-only, but a new safe-sex provision had just been added: "If you do have condomless sex it is assumed that you are on PrEP/Truvada or undetectable."

I wouldn't have noticed this email if it hadn't been for a response from New York's one remaining condom-only party. This wasn't an invitation but a statement of policy, an email unlike any sent previously or since. Safe sex is an important practice, it argued, and condoms are the only way to be safe. The second party remains condoms-only, and is still alone in this decision. It feels now like a holdover, a memory from a different time.

I've been to both these parties, the former in a dingy Brooklyn basement, the latter in a midtown private home, many of the same men participating in both. At a party, even if I'm not there to fuck, I can stand naked in a room full of other naked men, not a locker room where I'd have to hide the ways my eyes try to see the full extent of other bodies. Here, that's the point. Here, our hardness is exciting, full of possibility, nothing to be ashamed of. That feeling, exhilarating, even—I think—healing.

I'd been to both of these parties with Wesley, my partner, where we mostly only fucked each other. It was different than fucking at home, something like being in a porn, something about the potential stimulation in the air. If we wanted, we could fuck others, too, a frisson hovering between anxiety and delight. With anyone but one another, with any touch riskier than a blow job, we used condoms.

Wesley and I were born only a year apart. Our generation of gay men came after the plague but before the pill. I knew that 50,000 people died in the United States in 1995. I was 13. I knew that sex killed, that no pleasure is ever free of worry, of death. The first thing I learned about sex was Kaposi's sarcoma lesions, gaunt 32-year-olds on TV.

HIV didn't just kill bodies. It killed a type of sex as well, a type of pleasure. It erased the possibility of my body and another meeting, for a moment, without my mortality there too, watching. Sex is this: another body, my body, my death, all naked for all to see. I knew about HIV and death before I knew I was gay. I knew that being gay might be deadly, and now I sleep with men.

HIV has never left me. I'm nostalgic for the pre-HIV era I never knew. Our image of those years is ambivalent: you could give head in abandoned buildings by the piers, but anything like a relationship seemed impossible to so many, the notion of gay marriage laughable. Gay people weren't often permitted relationships in a world so threatened by our bodies and how we use them. Now we can get married, but we've lost the notion of pleasure without worry. I worry. I only have unprotected sex when in a monogamous relationship. Even then, who knows? Everyone cheats, even straight people.

Those who lived through or were born into the 1980s became a generation afraid of love and sex. HIV-positive writer Paul Monette wrote, in his 1988 memoir *Borrowed Time*, "I was already riddled with guilt: none of this would be happening if I'd never had sex with strangers. I suppose I felt there was something innately shameful about dying of a venereal disease." Alexander Chee tells a story of New York then: men climbing onto their roofs at sunset to watch each other jerk off from a safe distance; you couldn't hurt what you didn't touch.

Even after HIV became treatable as a chronic illness, I still viewed it with fatalism. Being positive would make it harder—I always felt—to find love and trust and sex. I had reservations about dating someone who was HIV-positive; I knew that if I were positive, others would have the same reservations about me. If there were a pill for my worry, I would take it, a cure not for an infection of the body but for the traumatized mind. I would take this pill now, and I would never stop.

....

Some definitions: MSM: men who have sex with men, a term supposedly devoid of political or social overtones (unlike, say, "gay" or "queer"). MSM are the community most at risk for new HIV infections. PrEP: pre-exposure prophylaxis. PrEP is a pill taken by HIV-negative people to maintain their status. Truvada: the one pill currently approved for PrEP today, in 2016. It has been used as a component of HIV therapy since 2004, but was only approved for PrEP in HIV-negative patients in 2012. In this essay, Truvada refers to the drug used specifically by those who are HIV-negative, a shorthand that is almost universal in New York's gay community. Undetectable: An undetectable person is HIV-positive but controlling their infection with antiretrovirals. New research shows that their likelihood of communicating the virus is zero.

Truvada came on the market four years ago, in 2012, but prescriptions didn't take off until 2014. Gilead Sciences, the company that makes and sells Truvada, reported in 2016 that 80,000 to 90,000 people were on PrEP; 12,500 people in New York State alone have filled prescriptions, the overwhelming majority in New York City. The number of individuals starting PrEP has increased exponentially, rising fivefold in two years, from the end of 2013 to the end of 2015. A survey by New York City's Department of Health estimates that 29 percent of the city's MSM ages 18 to 40 are already on PrEP.

For one week, those numbers included me. I wasn't having sex with a stranger. I was having sex with my ex, Kaliq. He'd been in and out of my life after he'd gone away for a month and I'd found evidence he was cheating. A part of me had always known.

We'd been having unprotected sex for a year. I've always had unprotected sex with my boyfriends, a sign that we cared for each other, that we had built something like trust. I insisted on couples trips to the free clinic after three months of monogamy. I loved this man in part because sex with him seemed so free, so out of my control. I begged him not to put me at risk. I told him my body was in his hands. He looked me in the eye and said I could trust him. I did trust him. After I caught him cheating, we used condoms. I got tested. He said that he never had raw sex with anyone but me, that it was an intimacy I alone had earned. I believed he was telling the truth. I trusted that he cheated safely.

My ex wanted another chance. He wanted to have raw sex again. So one of us—I don't remember who—suggested PrEP. This was in 2013, when hardly anyone we knew knew about Truvada. PrEP and undetectable were not yet listed as safe-sex options in hookup apps. Some gay activists still called it a poison, a party drug. This pill offered the promise of bringing us back together. Truvada was more certain than his word.

He borrowed a bottle of PrEP from a friend. We swallowed that big, blue pill once a day.

A week later, without touching him much at all, I was on my way out of his life. PrEP made space for me to consider raw sex with him again, but I realized that it wasn't HIV that made him unsafe. Maybe for the first time in my life, I wasn't afraid of HIV. I was afraid of him. There was no cure for the damage we had done to each other. So I left, and I tried to stay gone.

....

Truvada is not one drug but two: emtricitabine and tenofovir disoproxil fumarate. HIV treatment pills contain multiple (usually three) antiretroviral drugs. This is because HIV mutates rapidly to become resistant to any one therapy. As an RNA virus, HIV is sloppy when it replicates, making mistakes better known as mutations. Still, the likelihood of a single virus simultaneously acquiring resistance to two or three drugs is the product of the individual probabilities, a number that approaches zero without ever reaching it.

People at high risk for HIV infection—sex workers, MSM, those with multiple partners—can take Truvada and have unprotected sex with little risk of contracting HIV. There is a 96 percent reduction in HIV transmission for those who take the drug four times a week. For daily use, the reduction is 99 percent. In one key study, none of the participants contracted HIV. Another option—post-exposure prophylaxis, or PEP—can be taken after a broken condom or risky sex. These drugs stop the virus from replicating before it manages to find and infect T cells. The virus never becomes a part of us.

In preventing the transmission of HIV, PrEP is at least as effective as condoms. Condoms reduce the risk of HIV transmission through anal sex by 70 percent with consistent and proper use. For men who don't use condoms consistently—and according to studies, most men don't—the difference in rates of HIV transmission between sex with and without condoms is not statistically significant.

Those who remain HIV-negative while on PrEP will have antiretroviral medicine consistently in their bloodstream and no virus in their blood. Undetectable people have antiretroviral medicine in their bloodstream and no virus in their blood. In terms of HIV transmission, there is no reasonable distinction between those who are HIV-negative and on PrEP and those who are HIV-positive and undetectable.

For many years there was a respectability politics associated with condoms. Truvada was vociferously opposed by traditional gay health organizations. Michael Weinstein, the head of the AIDS Healthcare Foundation, continues to campaign against it. The opposition to Truvada seems to have made Gilead cautious about marketing it. Though the drug has been available for four years, Gilead has started to underwrite advertising only in 2016. Paranoia about PrEP remains, often driven by the idea that other infections (chlamydia, syphilis) will rise without condom use. Bacterial sexually transmissible infections (STIs) did increase in 2015, though it's impossible to connect that increase to Truvada. In effect, much of the rhetoric about STIs continues a long history of pathologizing gay sex, particularly raw gay sex, now that we can no longer rely on HIV alone.

When single, I use condoms consistently. I believed this: responsible, self-loving, caring, good gay men use them, always. I wanted to be that type of man. I was shocked, talking to my straight friends, to learn that they had unprotected casual sex. Gay men were considered unsanitary even before HIV. We have to constantly prove to the world that we aren't, in fact, sexual monsters, deviants. Straight people don't carry the same burden of politics, the same history of HIV, into the bedroom.

In the era of HIV activism, gay sex was central to the conversation. HIV was a sexually transmitted disease; how we fucked was how we lived or died. In the movie *How to Survive a Plague*, we see Peter Staley fighting with Pat Buchanan; according to Peter, we had to teach gay men how to use condoms. Human beings will never abstain from sex. We need to teach people how to have sex more safely! It will save lives! In the fight for gay marriage, we willingly hid our sexuality. An ex-partner of mine worked at GLAAD, a queer media training camp and watchdog, where he taught people to say "gay" and not "homosexual," because the latter puts the word "sex" in people's faces. The fight would be easier if people didn't imagine the icky things we do in the dark. We won the right to marriage by convincing straight nuclear families that our love is just like theirs. Our sex, too: three times a month, monogamous, missionary, 7 minutes, safe.

But there's always been tension. Queer people are a sexual and cultural vanguard. Anal sex is kind of our thing; now sitcoms joke about straight couples pegging. We had Grindr for years before Tinder popped up. We've been doing monogamish—being partnered but also exploring sex and sexuality with others both alone and together—since basically forever.

Gay and HIV activists fought conservative institutions—public schools, the Catholic Church—to make condoms widely available and to train people how to use them. In a world where everyone was dying, condoms were the only way to stay alive. They worked. They never worked well enough. For decades, condom culture was a type of care among gay men. When will queer pleasure *not* be under attack? But condoms may have a cultural significance that now surpasses their usefulness in public health and public policy.

Condoms alone fail to control the HIV epidemic, which is—even now—growing. People grow tired of behavioral interventions, like condoms; raw sex just feels better, and—over the long course of a life—using a condom every single time we fuck may be an impossible task.

I bought into the politics that binds HIV and gay-marriage activism: that condoms matter, that sex must be contained, safe, respectable. I was raised Catholic and always had been a little afraid of my own sexuality. Even when I lusted only after women, I viewed that lust as a sin. In my hometown, we waited until marriage. Being a virgin was *cool*, especially if you were a boy. As an adult, these conservative politics proved hard to dismantle. I thought myself better-than because I'd always used condoms. I looked down on my friends who didn't. Of course humans—of all sexualities—slip up. Of course people are going to find pleasure in doing the very thing they've always been told not to do. Sometimes we want things from love, from sex, precisely because they aren't safe. I wanted my ex; he wasn't safe. I always suspected he was cheating; we had unprotected sex anyway.

The conventional narrative of the past three decades is one of survival. We survived the plague. This story is flat, and in its flatness, untrue: the plague never really ended. But then we got marriage, an assimilation into an institution that I always found too narrow for most relationships, even straight ones. And what we lost was the freedom of queer sex, queer sluttiness, queer rage, raw sex, queer separatism, hedonism, and free queer love, which might not look like straight love at all. HIV gave gay men who believed in respectability, modesty, and monogamy the upper hand. Marriage used that respectability to gain legal rights. Truvada might be a step toward a new sexual liberation—sex parties, singles and swingers, threesomes even for committed couples—and away from the condoms that made our sex safer not just physically but culturally. No wonder PrEP makes people, gay and straight alike, uncomfortable.

....

Truvada is nothing special, nothing new. The antiretrovirals in it have been used for decades. The difference is the bodies the drugs are put inside, now HIV-negative, no virus in residence, now not ill but pre-ill, infected only by the type of sex we have.

And yes, it is also about profit. The pill's out-of-pocket cost is roughly $18,000 a year. Truvada is now made by only one company, Gilead. One common criticism of PrEP is that it requires HIV-negative people to take very expensive pills whose side effects are not insignificant. PrEP, the argument goes, turns gay sex into a profitable (and therefore palatable) enterprise in the age of late capitalism, where everything is moral if it's making someone rich. Gay sex parties aren't sinful debauchery; they're added value for Gilead shareholders.

The rebuttal is that PrEP works. It's most likely less expensive, and involves fewer years of dealing with side effects, than taking antiretrovirals for a lifetime, as those who are HIV-positive must. Yet $1,500 a month for the option of sex uncontaminated by fear of HIV is a high cost for an individual or society to pay.

Gilead has a program that provides free drugs to those without insurance, and there is a co-pay assistance program as well. Theoretically, anybody should be able to get Truvada at low cost. In practice, it's not that simple.

As I was writing this essay, I had dinner with a friend who had been on Truvada but who had recently had to stop taking the drug. He hit the yearly cap—$3,600—for the Gilead co-pay assistance program and his co-pay, he said, was as high as $500 a month. With private insurance and a very high co-pay, he was—in terms of access to Truvada—in a worse position than those who have no insurance at all.

"I just want to be able to have sex again," my friend told me. He's single and mostly uses online apps for both sex and dates. He was having a hard time finding men who would have sex with condoms. For my friend, in 2016, PrEP feels necessary to have sex. He wouldn't have unprotected sex without it, and hookup culture—according to him—has moved on. Condoms are no longer the norm. This does feel new.

I knew the city was building a program to help people with private insurance who max out the Gilead co-pay program. I had met with Dr. Demetre Daskalakis, the head of New York City's HIV prevention campaign, to talk about condoms and PrEP and the city's programs. We'd discussed people in my friend's situation. I wanted to help my friend, or see what he'd have to go through to get the drugs he needed uninterrupted. I spent the next two days on the phone with a full alphabet of agencies: first 311, then HASA (only for the HIV-positive) and ADAP (doesn't cover PrEP), then my city-associated HIV clinic. Nothing. Then state programs, PrEP-AP and the PrEP hotline; then PAN, then Xubex.

On hold listening to badly performed classical music I thought, *This is how bureaucracy kills: to Bach.* I'd spent months researching PrEP, and I couldn't help my friend. Once I'd exhausted all the options, I reached back to Dr. Daskalakis, hoping for a nudge in the right direction. Yes, the city knows about the PrEP donut hole for the underinsured. Yes, they are hoping to develop a solution. No, there is nothing yet. It's been months since my friend has taken Truvada and months, too, since he's had sex.

Truvada offers sex without worry for $50 a pill. People live and die on the basis of a brief conversation at a clinic about whether and how to sign up. In some places people are still—against medical and epidemiological evidence—going to jail for fucking while HIV-positive. Many still blame an HIV-positive person for their risk when both partners consent to unprotected sex. Even in New York, where the city government is committed to PrEP and stigma-free HIV policy, there are countless people who fall through the cracks. Outside of this city, those with the least access to PrEP will be people—queer people, poor people, people of color, people in prison, people in rural towns—who have always been excluded from healthcare.

The coming years could make geography matter even more than it already does: drugs like PrEP may be accessible in New York and entirely unavailable in rural America. For years now, needle exchange programs, which prevent HIV transmission, are being attacked—including in Indiana and West Virginia. More than 500,000 people in America know they're HIV-positive but aren't being treated. This is more than half of all people living with HIV. A friend of mine—Jesse, the gay scientist who lives in DC—has a cousin in a deep-red state dying of AIDS, right now, today. He has KS lesions, he's unlikely to live much longer, he often can't get drugs, and when he does take medicine, he tells his relatives only that he has "cancer." This isn't the story we're telling ourselves, and it's not one we often hear.

....

When PrEP was introduced, we all wanted to know: Would it change behavior? Would it lead to a world where condom use dwindled and raw sex became—again—the norm? Early research showed that PrEP users didn't stop using condoms. People on PrEP weren't more likely than they were before to have raw sex, they were just more protected in their actions.

More recent studies suggest otherwise. My experience and the experience of my friends also hints at a culture shift. Almost everyone I asked who still uses condoms has a story of someone backing out of a hookup if condoms were to be used. People are having more raw sex, and they're more open about it. They advertise it on apps. They talk about it with their friends. I see it now, and I didn't see it much before.

I've spoken to dozens of people, on and off the record, about PrEP. I was recently sitting at my kitchen table with three gay writers, Tommy and Denne Michelle and Fran, none of whom were taking PrEP. I realized that all the people I've talked to—whether they are taking the drug or not—are making a difficult and informed choice about their body, their pleasure, their risk, how they have sex. Some are using PrEP as a backup to condoms, others as a substitute. Many aren't using it at all; some use it on and off, when they are not in a monogamous relationship. Some want to start. At my kitchen table, Tommy said that having lived through the 1990s, he'll never take the condom off. It'll never—for him—feel safe or sexy.

Me? When my boyfriend Wesley read an early draft of this piece, he asked if I wanted to get on PrEP. "No," I said, "I don't. Do you?" He thought that because I was writing about this, I wanted raw sex outside of our relationship. No, I just wanted to understand the world I was seeing and living in, and I needed to write to do that. He didn't quite believe me; how many people hide their desires from their boyfriends, if not themselves, too.

This moment we're in makes me feel hopeful. Even if HIV happens, so what? I used to think that HIV would make it harder to find love and sex. Now we know that HIV-positive and undetectable is safe. It's sexy. I have friends who prefer to sleep with undetectable men. They know that most HIV transmission is by people who don't know they're infected or aren't on treatment. For an acquaintance who prefers not to use condoms for his hookups, people who know they're positive and are on drugs are the safest bet.

PrEP and U=U—the idea that undetectable people cannot transmit HIV—did not change the virus, the material that it's made of, what it looks like. They changed our relationship to it. This is the power biomedicine has. A power that's too often available to those who can afford it.

So yes, we now have good HIV medicine for both treatment and prevention. Pills don't cure us, but they might keep us alive. In 1996, people were so near death, and then they weren't. Pills do beautiful, beautiful things. But pills can't do it all. Pills can't make us better at negotiating consent or understanding risk. Pills are not healthcare infrastructure in communities that need it. Pills don't erase stigma. Pills exist where people can afford them. Pills exist for people who can get to doctors, clinics, and hospitals.

In New York City, Truvada has become somewhat mainstream. All living is risky. All sex, too. I've slept with people I didn't love enough when they loved me deeply. I've slept with people I loved who didn't love me enough, who lied, who cheated. I've had joy, too, even with that ex: when we made love as midnight brought in my thirtieth birthday. Joy: the Grindr hookup I had whose body fit mine. Joy: the first kiss with my Wesley, my lover, my roommate, my boyfriend, a bundle of nerves, leaning forward on my couch, our glasses clinking at the nose. We met on Grindr, neither of us on Truvada. We'd just planned on kissing, maybe a blow job, but I liked him and he liked me, and so I pulled him into my bedroom and he fucked me, safe with a condom. It didn't break. For decades, and still too often, these small moments of pleasure could bring death. Even now, a third of the people who might most need PrEP don't have access. Even now, I know that my ability to heal from my lifelong worry comes at a cost because it's not available to all.

In the past three years, I've been able to imagine a new type of pleasure. Remarkably, this pleasure is one willing to inhabit my own body. I don't know if it's PrEP—even though I don't take it—or the idea that being undetectable is safe, healthy. I don't know if it's due to a partner I trust with my life, but I suspect it's something more than that.

Even with boyfriends or girlfriends I trusted before, I could never have sex without feeling my life and death were at stake.

Even with Wesley too, at first. We didn't change. I haven't changed. The culture around us changed, and this change infected us. I count us among the lucky, and we all deserve this luck. With Wesley, we stripped each other naked, no Truvada in either of our blood. In these moments, my mortality shrank. Sometimes I didn't think about it at all.

I was right to trust this man with my life, but that didn't mean he wouldn't leave. Only weeks after this essay was originally published, when he did dump me for a job abroad, I went to the doctor and asked for a large blue pill, daily. The safe intimacy between my boyfriend and I broke. It was time for me to make safe intimacy, safer sex, all on my own. I'd rather have had him, but he was gone. Everyone deserves pleasure without worry, at least for a moment, at least for a fuck, even us fags. Even me. I wasn't taking a pill in order to accept a man who cheated. I was taking it to make safe pleasure all on my own, which I never could have imagined. Everything was wrong about coming home to that apartment empty of the man and the dog I'd come to love so dearly, love like my own flesh. At least I didn't have to suffer that particular viral fear any longer. No. No, that was alright. It would be. I took a big breath, and—to move on from all that, from all this—I swallowed.

6

On War

The Limits of American Moral Reasoning

WITH PATRICK NATHAN

No one loses a war like the United States of America. Yet we start them all the time. In fact, if you pause to listen to our language from "battleground" states in election years to "culture wars" every other decade—it's hard to believe our nation is not at constant, unending war with everyone and everything, including itself.

This is, in part, because the United States of America has no idea what a war is, nor how, any longer, to wage one. In fact, the last time Congress officially declared war—the only way, it should be said, that the Constitution allows for war—was in 1942, against Bulgaria, Hungary, and Romania. Since then, militaristic violence overseas has been "engagement" or "intervention" or "occupation," and most of our "wars" are waged not against nations but against concepts or abstractions like terrorism, poverty, and drugs.

Or viruses. On March 28, 2020, as New York City's hospitals began to fill to capacity, President Trump tweeted: "With the courage of our doctors and nurses, with the skill of our scientists and innovators, with the determination of the American People, and with the grace of God, WE WILL WIN THIS WAR. When we achieve this victory, we will emerge stronger and more united than ever before!" Six weeks later, he told the press that healthcare workers going through the doors of hospitals were "running into death just like soldiers run into bullets. It's incredible to see. It's a beautiful thing to see."

The virus, of course, was not his only "battlefront." In late May of the same year, when the murder of George Floyd dominated the news cycle and protesters marched masked on the streets of hundreds of cities, Trump's defense secretary Mark Esper encouraged state governments to "dominate the battlespace" as Trump himself threatened to use the 1807 Insurrection Act to use the military against peaceful protests. Around this same time, a sitting US senator, Tom Cotton, wrote an op-ed for the *New York Times* entitled "Send in the Troops."

Two months later, as school reopenings loomed, Donald Trump's all-out "war" against COVID-19 had been, like all other American "wars" since 1942, an abject failure: we led the world not only in defense spending but in COVID-19 cases and deaths (and inequality, and domestic terrorist attacks, and opioid addiction, and . . .). Throughout the United States, people—usually the former president's people—continue to reject social distancing and mask wearing while spreading scientifically unsound misinformation and conspiracy theories. In Georgia, Governor Brian Kemp, a white Trump-aligned Republican, banned mask mandates in his counties and cities, going so far as to sue Atlanta mayor Keisha Lance Bottoms, a Democratic woman of color, for insisting that city residents mask themselves.

This was a "war" we were destined to lose. Not because we don't have the resources necessary to implement the nonpharmaceutical interventions (distancing, mask use, hand hygiene) that we know work, as exemplified in countries like New Zealand and Japan. Wars are won through mass death. A virus will never be dominated.

COVID-19 couldn't then be beat. Now, vaccines are a tool that could, if we wanted, prevent more suffering from this novel virus. But around the world, vaccination is still largely stalled. Not vaccinating the world remains a choice, one the United States under Biden seems intent on continuing to choose every day, even if that choice is passive, a lack of action. The tools of care, whether they're a mask or a shot, seem to matter little when it comes to implementation; we seem unwilling to act, placing a false notion of personal freedom—*I* don't want the shot! *I* won't wear a mask!—above a simple and safe way to care for ourselves and one another. A new rhetoric of care, empathy, and respect for life is needed to face COVID-19 and survive—a rhetoric of care encoded not only in how we speak but *also* in the structures and infrastructures of our nation, from our healthcare system to our need to dominate by overpolicing, to our economy, our military, our education.

Even when we want to help or want to survive, it is too easy to let the language of war occupy (see?) our tongues, to use metaphors of death making to speak of a virus and our body's immune response, or a virus and our nation's collective response.

War has failed us, and because of that, we have failed the world. What America needs now is care, not war—spoken in words, enacted in policy.

....

"Your health" is what we drink to—and who wouldn't want it? To eat well, to exercise, to be thin and fit and grow old gracefully before dying out of sight. To feel proud when a nurse calls your blood pressure "perfect" or your body fat percentage "normal." To smile when they flick the needle and say, "Such healthy veins."

This assumption of health as an "identity" is one Eula Biss

reflects on in *On Immunity*, her 2015 book about vaccination and togetherness: "*I am healthy*, we tell each other, meaning that we eat certain foods and avoid others, that we exercise and do not smoke. Health, it is implied, is the reward for living the way we live, and lifestyle is its own variety of immunity."

It is virtuous, then, to be healthy, just as it's virtuous, in America, to be rich. The healthy and rich, it is understood, have worked hard to get where they are. And now—in the times of a global pandemic—their bodies are among the safest. Many white-collar workers are able to work from home and have their groceries delivered. Those contracted (not hired) by Whole Foods or Instacart and other food purveyors to deliver these groceries during the "lockdown" in 2020 did not have this option. Many from that same work-from-home class began venturing out into bars and restaurants, choosing to shed the mask for an hour in a public space in exchange for a meal someone else has cooked. The workers who wait on them—and who are disproportionately Black and Brown—have no other viable options: they can't afford to stay home and safe. Much of this health risk, and who bears it, is grounded in our society's easy conflation of health and financial privilege.

This kind of virtue, flatly earned, can be stifling. *If I'm a good person because I'm healthy, what happens when I'm not?* Over these long months, as most of us self-isolated at home or with small groups of friends and family, our typical rhythms of keeping our bodies well may have been entirely disrupted. We ate only what's available. If gyms *were* open, they were among the riskiest places to go. Going on walks and runs and bike rides carries an aspect of shame never present before, as getting out of the house was being reported—and often uploaded—as singly selfish behavior.

Illness has long been blamed on the ill. Even cancer—once considered a disease of "repressed feelings" but now understood as a consequence of genes, certain viruses, and environmental factors called carcinogens—still carries a hint of its former moral condemnation. A healthy person *should know* to avoid carcinogens. To get cancer in a society where health is part of one's lifestyle is a sign of failure, a punishment for laziness or a lapse in strength. In the popular imagination, what cancer often means is that, somewhere in life, you fucked up.

Demanding that sort of health virtue from others is a way to render those who don't measure up inhuman, incapable of agency. This is fundamental, as Elaine Scarry wrote in *The Body in Pain*, for the imagination of war. Individual soldiers, whose bodies will be injured or destroyed, are rendered invisible: "The overall army or overall population, and not the fate of single individuals, will determine the [war's] outcome." Personifying the army—"the imaginary body of a colossus"—is a "convention which assists the disappearance of the human body from accounts of the very event that is the most *radically embodying* event in which human beings ever collectively participate."

The citizen's body, Scarry writes, "may in war be permanently loaned in injured and amputated limbs. That the adult human being cannot ordinarily without his consent be physically 'altered' by the verbal imposition of any new political philosophy makes all the more remarkable [that he] agrees to permit this radical self-alteration to his body." War "requires both the reciprocal infliction of massive injury and the eventual disowning of the injury so that its attributes can be transferred elsewhere." Leaders will say the conflict wasn't about inflicting pain or killing millions of people; it was about freedom, justice, unity. The nation. The national body. The individual human beings sent into battle were not sent, leaders will say, "to alter (to burn, to blast, to shell, to cut) human tissue"; they were sent to defend democracy or "our way of life." The real subject of war—an incomprehensible scale of injury inflicted upon the human body and the places it calls home—not only disappears in the language of political or moral philosophy, but is meant to disappear.

This martial erasure is echoed by the politicians and pundits who argue that it is virtuous to reopen the "economy," or the restaurants and bars and schools and universities that keep their constituents busy. Opening the economy protects the "American way of life," even if individual lives are lost (which never means theirs, of course).

With COVID-19, the language of war is clear. And it's not just Donald Trump. To quote the *New York Times*: Our country is "*armed* with some of the most highly trained scientists and infectious disease specialists." Paramedics in New York call their hospital "a war zone"; their EMTs are "on the front line." One op-ed writer, Elliot Ackerman, compares the threatened "70 percent"—or the "proportion of Americans ultimately infected with the virus"—to "the estimated casualty rate of killed or wounded for my platoon as we prepared for the assault into Falluja, the largest battle of the Iraq war."

Even scientists obscure infectious diseases and our immune response with metaphors of war. "Cytokine storm," the immune overreaction to the virus that can lead to death, sounds like some general's pet military operation. A paper on drug development defines the efficacy of "chloroquine and hydroxychloroquine as available weapons to fight COVID-19."

In this "war" against COVID-19, we can also see how society suspends the agency of exposed "essential workers" by constructing them as "heroes." Doctors, baristas, grocery store clerks are all transformed into a "they" whose self-sacrifice saves "us." With no one obligated to protect them, these heroes—like infantry soldiers in a war of attrition—can be dehumanized, silenced, and invoked, alive or dead, in bad-faith political cant. Refused the right to imagine themselves as "healthy," doctors, nurses, and healthcare workers don't need PPE—they're heroes for working without it! Grocery store clerks do not need a living wage and health insurance—can't they hear us clapping and banging pots and pans from the balconies of our fourth-floor walk-ups? Amazon workers don't need sick leave; they're just lucky to have a job at all—and it's certainly no time to be organizing, unionizing, fighting for worker's rights, or walking off the job! As the Raleigh, North Carolina, police department put it in an April tweet: "Protesting is a non-essential activity." Don't those workers know we're at war?

There is no "choosing to be a hero" in continuing to work for wages you can't afford to refuse, yet we're meant to applaud rather than rage when people get sick and die. Their bodies risk infection with a potentially lethal virus, yet we are not meant to think of their bodies—only their "sacrifice." If they are "soldiers," their deaths are just part of the job.

There is not any virtue in having a healthy body. Disease is part of the human experience. Viruses have coevolved with us, buried in DNA we acquired long before we became ourselves. We carry snippets of the Neanderthal genome in ours—and why? We hypothesize that these sequences protect us from diseases that existed in the eons we still interbred with those more ancient hominids.

And yes, disease changes us—viral infections especially. The human body evolved a molecular memory that protects us against reinfection. And yes, the molecules of our immune system pressure the virus to evolve; and so we alter the virus as well. As it moves from one cell to another, from one body into another, it mutates to evade the immune system, and the body learns to again recognize it. As Biss wrote of her infant's immune system, it is not unarmed or defenseless but "what immunologists call 'naive.' . . . With or without vaccination, the first years of a child's life are a time of rapid education on immunity—all the runny noses and fevers of those years are the symptoms of a system learning the microbial lexicon." This is not a war at all but a conversation, one held over weeks, months, years—the lifetimes of people, yes, but also of our species and the viruses we will always live alongside.

To resist the militaristic framework would be to acknowledge this conversation, to engage with it. It would be to acknowledge the horrific "peacetime" conditions of American workers, essential or not, as well as our underfunded and outdated public health system—ever shrinking, like every other nonmilitaristic aspect of the federal government, against the vastness of military spending. To see outside the limiting, false parameters of war, we need to reimagine our quarantine and ongoing efforts at social distancing—right down to their language—not as detention or sacrifice but as community acts of care.

If we *are* to imagine ourselves as a national body—even a global body—this is a virus we need to listen to, not fight. The fractures it illuminates in our culture and society were preexisting conditions. The immunity this virus may leave behind in our blood could become a cultural immunity as well, a society more robust to crisis, but only if we learn the lessons placed now before us.

....

Here's a different metaphor: Like viruses and the human genome, diseases—particularly the nasty ones—get assimilated into our cultural DNA. Tuberculosis, as Susan Sontag pointed out in *Illness as Metaphor*, changed how the public perceived artists and other "sensitive" individuals, as well as providing the blueprint for the contemporary fashion model: thin, sickly, pale with flushed cheeks, "reduced" to nothing but youth and vivacity. In a more obvious way, HIV has altered our relationship, worldwide, with sexual pleasure, as well as leaving a lasting, isolating stigma on the queer community. As for COVID-19, it will inevitably alter a generation's experience of social interaction and physical intimacy.

In 1878, the young and vaguely aristocratic Wilhelm von Gloeden moved from Germany to Sicily to "take in the air." He'd been diagnosed with tuberculosis and prescribed the usual treatment—high places, dry places, sparsely populated countrysides with little opportunities for agitation. While in Sicily, he began photographing not only the landscape, which helped to promote British and northern European tourism of the island, but also the island's boys and young men, many of whom he took (in perhaps both senses of the word, where the younger ones were concerned) as lovers.

Even by today's standards, von Gloeden would be considered a criminal "other" or outsider; and certainly in the nineteenth and early twentieth centuries, he was a deviant who went against the social contract. Yet, isolated as he was, he was tolerated, his work purchased in salons across Europe. It wasn't until after his death—when Mussolini's Fascist party, allied with the Vatican, confiscated and destroyed more than half his work—that "society" concerned itself with von Gloeden's morality. Before that, he was too remote—perhaps too far away, on his island of bronze boys, from the bourgeois sons of London and Paris and Berlin—to be of any concern.

Tuberculosis is a disease of the lungs caused by a mycobacterium. Like all bacterial and viral diseases, it has nothing to do with one's personality, politics, or ethics. Yet as Sontag points out, this was not, when TB was still untreatable, the going belief.

What she doesn't suggest is that perhaps TB became so overstuffed with metaphorical thinking and moralistic psychology because treatment for the TB patient was removal from society, an exemption not only from labor but from family life and imposed social interaction. The TB patient, wandering from the desert to the mountains to the Mediterranean coasts, from sanatorium to sanatorium, was an economic and social subversive, liberated from the roles capitalism assigns to the rest of us. Their isolation wasn't just about them, but about *our* protection from *their* disease.

The best way to ensure that their extra-capital lifestyle did not become a threat—did not become attractive or desirable—was to cast them morally apart. The psychological TB archetype is capitalism's way of protecting itself from what it has always seen as a threat: that disease, infecting a wide variety of victims, may reveal our social connections. *They* may infect *us*. Once the bacillus and its antibacterial medication were discovered, the TB patient was no longer exiled, but treated within, and assimilated by, the same socioeconomic and biomedical framework as the rest of us. They were no longer morally diseased, only physically sick.

Today, endemic TB cases follow a familiar and effective social response: we test, trace, treat, and quarantine. We do this because TB is not personal. No disease is personal. Even noninfectious diseases like cancer or multiple sclerosis, whose etiologies are still not wholly understood, are not the victim's "fault." Illness is accident, not malice. It is amoral, not immoral. It's one of our universal truths as humans that we all get sick.

Treatment and prevention, however, *are* personal—at least in countries with personalized healthcare like the United States. The sequencing of the human genome birthed an era of "personalized medicine," wherein our therapeutic responses to any illness, from cancer to infection, could be tailored specifically to our unique and individual genetic background. Of course, this promise remains ever elusive as our individual genome sequences have very little to say about how to treat any known disease, and of course it's all so complicated, and most of it isn't predetermined but horrifyingly random.

Our likelihood of dealing with our random, impersonal illnesses depends on our ability to care for ourselves, which makes illness a marker of class, race, and power. It makes socioeconomic status all the more visible: Are you insured? Do you have a primary care provider? Do you have the money for co-pays and tests? Can you afford to stay home for weeks as you recover? Rather than betray you as *human* and equal to your neighbors, disease in America can betray you as rich or poor, coddled or neglected, white or Black, "healthy" or "unhealthy." According to a 2009 study, 60 percent of bankruptcies in the United States are due, at least in part, to medical debt. We are always and already losing lives and livelihoods to the profit-driven system of American health "care," where you only receive care if you can afford to.

We're all given bodies that can and will break down. During a pandemic, the risk to the body is even more *visibly* impersonal. Scientists estimate that to attain herd immunity against COVID-19 infection, 70 percent of our population would need to be infected and subsequently immune. The only thing 70 percent of people in this country have in common is being here and being alive. Unlike diabetes, heart disease, HIV, and countless other illnesses, contracting COVID-19 is not necessarily a *visible* marker of class (though working-class people *are* more likely to be exposed to it, as they are placed in the path of the virus). But *dying* of COVID-19 is. What determines these two facts? Generations of health disparities, of structural violence, of the legacy of chattel slavery and migrations that are still encoded in law, science, and health. It is not, as Trump's doctor (our former surgeon general) suggested, because Black Americans smoke and drink too much. It's because of a 400-year-long project of using Black and Brown Americans as guinea pigs and then setting up systems of care that make the technologies tested on them harder to access once they're deemed safe enough for the rest of us. Ask Henrietta Lacks about the "war on cancer."

Health remains, in its severity as a public phenomenon, a visible marker of a society unwilling to care for us all. It's another "identity" imposed on us by a political system that pretends to be tired of "identity politics." It's not a moral failing to get sick, but it is a society's moral failing to refuse social accountability for a social phenomenon like illness.

Statistics in times of plague are cruel. A small percentage of COVID-19 patients, no matter their age or history of health, will develop severe disease. When only a thousand people are infected—as was the case in the United States early in February—this seems to be a disease of older, already vulnerable Americans. But multiply those risks over millions, and we understand how connected all our bodies really are. Late in 2020, we learned that the excess mortality to young people, ages 25 to 44, between January and October was nearly 40,000 people. Young people die too. We knew they would. It's a question of simple math, statistics. Rare events become certain. With millions vaccinated, even with a near perfect vaccine, breakthrough infections are mathematically guaranteed. They don't speak to a vaccine's failure; they're proof of its success at the scale of (almost) nationwide use. COVID-19 will touch all our lives; each of us will lose at least one person we know, but some of us—those in what the state calls "unhealthy" communities—will lose many, many more.

No *body* is safe from COVID-19. Rather than render visible any individual body's failure to ward off illness, this illness renders visible our entire nation's failure to care for all of those who are ill.

....

The imagination of war is an imagination of failure—indeed, of total destruction. As a "foreign agent" originating in Wuhan, the virus quickly triggered the former president's predictably racist, xenophobic coping mechanism—one of banned flights, sealed borders, and impenetrable walls to make an "airtight," sterilized country. Thanks to him, the United States now leads all other nations in infections and deaths. While the CDC knew, even in January 2020, that COVID-19 could be transmitted by people who did not feel ill or exhibit symptoms, the president imagined he could magically halt the virus at the border and—like other "infectious agents" he's rhetoricized—deport it. This, and not a functional public health infrastructure, would keep our national body intact.

Like all of this country's endless, doomed wars—on terror, on poverty, on drugs, on cancer—there is no "fight" against an invisible, insidiously personified *yet crucially nonhuman* enemy. There is only acknowledgment, alleviation, and treatment.

War is a favored metaphor in this country—"a society," as Sontag wrote, "in which it is thought foolish not to subject one's actions to the calculus of self-interest and profitability. War-making is one of the few activities that people are not supposed to view 'realistically'; that is, with an eye to expense and practical outcome." In America, war is perhaps the lone call to social mobilization that rarely earns the politician's or the pragmatist's groan "But how do we pay for it?" Where nothing escapes commodification and where human beings, even in the midst of a pandemic, are appraised and ranked by their economic value, war is transcendent—the only "greater good" our representatives are able to predictably sell back to us. Even torture, as the second Bush administration proved, is a heroic act of sacrifice.

When applied to illness in particular, the military metaphor, Sontag continues, "implements the way particularly dreaded diseases are envisaged as an alien 'other,' as enemies are in modern war; and the move from the demonization of the illness to the attribute of fault to the patient is an inevitable one." While immoral in its actions—causing mass injury to others—war is moralizing in its imaginative parameters: there is an "us" to represent the good and a "them" to represent the bad; and the line between the two is meant to be clear, visible. Needless to say, it's always the state that draws the line, the state that pulls the trigger. In reality, only nations are legally able to be at war with one another. When we wage "war" against unseen, personified concepts (drugs, poverty, terrorism, HIV, the coronavirus), we are forced to imagine a perpetrator. All wars are wars against. And who, in these cases, do we imagine? Drug addicts, the poor, religious extremists radicalized by unstoppable American violence, HIV-positive individuals, and so on. After all, it was in response to the War on Poverty that Reagan's racist, classist straw man of the Welfare Queen was used to eliminate the social safety nets of millions. The "War on HIV" built the stigma against HIV-positive individuals that drove bad science for decades until we understood, finally, that undetectable equals untransmittable, that HIV-positive people are indeed *safe* to sleep with.

With COVID-19, the most obvious moralization inflects hygiene, social distancing, and race. To be sure, hygiene and social distancing practices can not only save one's own life, but the lives of one's neighbors. It's tempting to want to use whatever megaphone one can, be it Twitter or screaming from your own front porch, to shame people into going back inside.

But reimagined in metaphors of war, those who fail to practice social distancing become traitors—"they" are against "us" in the effort to eradicate the spread of the virus. The July 4 circuit queens of Fire Island, for example, "joined" the virus as agents of contamination by partying, hundreds deep, without masks. This lends, then, a militaristic view of quarantine itself: confinement against one's will for the benefit of others—a double offense against the peacetime American idyll of roaming wherever you please and not feeling obligated to care about others while you do it.

But even more visibly, the racist component of this war metaphor has driven attacks against Asian Americans as well as Asians worldwide; they are imagined as "invaders" and face a similar violence that queer individuals or effeminate men faced in the 1980s and '90s as AIDS killed millions of people worldwide. To be labeled "contagious" is to be implicated in the activities, the intelligence, of the virus: one's patriotism dragged again before public scrutiny. An infectious identity—HIV—led to anti-queer and anti-Black violence in the 1980s; to be queer was to be a threat to the "rest of us." William F. Buckley suggested—in the pages of the *New York Times*—forcibly branding those with HIV to protect this uninfected rest. And who exactly would be branded but the four Hs: homosexuals, heroin users, hemophiliacs, and Haitians.

From HIV to COVID-19, then, public metaphors around the origins of pandemics and the people most likely to get ill and transmit the virus do racialized harm to people with fists and knives, not infections. And these stories, of viral introductions into humans, are simplified into falsehood. HIV comes from the Congo, yes, but emerged in the aftermath of colonization and its urbanization and into a population of Congolese who were not allowed to become doctors because only white Belgians could do that. COVID-19 comes from a part of the world where urbanization leads to increased interaction between humans and animals, like places on all continents where viruses can spill over into humans. The narratives we map onto already vulnerable people when viruses spread are false and dangerous. The murderer who killed eight in a rampage of spas run by Asian and Asian American women imagined stories for these women that included only the pleasure they offered him; to hear their families speak about these women shows how false his imagination was, and how deadly. The white imagination seems always to need a neat narrative of blame; the complex world we live in so often offers nothing of the sort.

In 2020 and 2021, we saw this logic perversely turned on Asian Americans and Asian immigrants, even as Asian countries rapidly brought their COVID-19 epidemics under control. Racism and homophobia do not follow the rules of logic, but of power-over and brutality.

Now, after months of self-isolation have faded into partisan displays of patriotism—is that mask protecting the country, or are you "against Trump" for wearing it?—the visibility is one simply of reminders. A mask—the most effective, easily affordable tool alongside vaccines for preventing the spread of respiratory viruses while still existing in social spaces—has become a political signifier: those who wear them "believe in the virus" and therefore (it is imagined) support its presence in daily American life, while those who shun the mask—and attack those who ask them to please wear one—are "resisting the tyranny" of the virus, not to mention the tyranny of all efforts to care for those who are ill and prevent many more millions from becoming sick.

All metaphors aside, social distancing *is* effective. In Wuhan, new cases went from nearly 3,000 a day in February 2020 to fewer than 100 a day in March with aggressive testing, treatment, and assurance that people—infected and not—stayed home. In South Korea, more than 30,000 people were quarantined in March, and despite an initial surge in infections, the curve rapidly flattened. Even in certain US states where quarantine measures were adopted, clearly communicated, and taken seriously—New York, for example, or Washington—the rate of infection slowed significantly.

While often viewed as militaristic and enforced isolation, quarantine and social distancing can be reimagined as community care—deeply social activities that don't lock us all behind barbed wire but remind us, even when we're alone, just how much we rely on one another to live healthy, safe lives. From the beginning, we've known that COVID-19 is most dangerous for our elders and for those who are immunocompromised. We can also see quite clearly that, in America, it's more dangerous for Black and Brown and poor people. There is here, too, a surge in the number of young and ostensibly healthy people dying. While those at lower risk for severe disease might be *inconvenienced* by the virus, they can carry it to those who might be incapacitated, hospitalized, or killed. And those at lower risk for severe disease within a week or two of being infected by SARS-CoV-2 might end up with a debilitating and chronic reaction to the virus, so-called long COVID—now called post-acute COVID-19 syndrome, or PACS. Viruses, and our body's reaction to them, defy expectations and logic. Sometimes mild COVID never ever goes away, the seemingly mild symptoms not at all mild when they last for months or maybe years. I wouldn't want anyone to suffer this way. Caring for others through simple-enough measures (wearing masks, getting vaccinated, distancing) is the only way to prevent that suffering.

At the same time, given COVID-19's high rate of infection (5.7 new infections can be expected from every infected person), isolation measures ensure that hospitals are not overwhelmed with seriously ill patients.

In quarantine, you might be all by yourself, but through that act, you prove that you know you're not alone—not in your community, and not on this planet. Quarantine is a social act, not a personal sacrifice.

....

The pandemic has complicated the autonomous image of the American—the person who does whatever they want. To be indulgently American is itself suddenly un-American. The imagination of war helps negotiate this loss—by obeying quarantine, you are *serving* your country; isolating yourself and limiting travel is a way to *fight* the virus. In this way, the war metaphor ensures that even our most selflessly social acts are assimilated into the realm of personal choice: it's still, we get to think, about us, about the real Americans.

This is a seductive way of thinking. In Ackerman's imagination, for example, the "war" against COVID-19 needs "wartime leadership"—shoes the former president, unsurprisingly, proved too small to fill: "Mr. Trump's initial Easter forecast bears a chilling resemblance to the hollow promises made at the onset of other conflicts, such as our Civil War and World War I, in which the troops would 'be home by Christmas.'" Obviously, the COVID-19 pandemic is not a conflict involving two nation-states, but Ackerman is right in pointing out Trump's failure to play along with the metaphor: he wasn't holding up his end of the bargain, unable "to deal with the American people honestly" and "be straight with us about the facts of this crisis while affirming that we're completely capable of rising to any challenge as a nation." Trump openly put the economy before "American lives," something no president is supposed to *admit* to doing. Ackerman channels three more conflicts: "From Sept. 11 to Pearl Harbor and back to our nation's founding, Americans know how to come together"—an erasure, needless to say, of Islamophobic hate crimes, the "quarantine" of Japanese Americans, and the Atlantic slave trade and the genocide of Indigenous peoples, respectively; but those presidents, at least, knew not to fret in the newspapers or on the airwaves over lost dollars and slumped markets. They knew how "total" the language of war has to be, even if war, too, is just another transaction our leaders make against the larger world. Instead of leading the American people, Trump was leading the American economy—the health of which he fret over constantly while neglecting real, living people.

But, as we've seen over waves in the past many months, mass death and uncontrolled viral infection are actually bad for the economy. The countries that have had minimal impact economically are exactly the ones who shut down entirely to stop the virus from spreading. South Korea only shut down indoor dining for a few weeks. Japan maintained an unemployment rate under 3 percent; ours ballooned to nearly 15 percent. Australia and New Zealand fared extremely well, although, as island nations, it is simpler to control people coming and going. So yes: keeping COVID-19 cases low did require controlling people's movement, which so many people (particularly in our country, it seems) consider their freedom, their right, to have, other peoples' health be damned!

Trump was a phenomenon the war metaphor did not anticipate. What is dying, in his narrative, is not a friend or a neighbor or even an American but something far more important to him and the party he leads: unfettered profit. He reveals the intellectual poverty of the war metaphor: it clearly did not change how America operates in an emergency; it only revealed how we've always operated. We are not *with*, and never have been—only *against*. Each other. Our neighbors. Our pasts. Ourselves. War, as we've already seen in this short but devastating war against a virus, never saves us. War only kills.

....

To go back to that metaphor about our "cultural DNA": Instead of "fighting" the virus that causes COVID-19, what if we were to learn from it? What if we were to imagine this pandemic as an opportunity for a societal education? This, of course, would "confer meaning," in Sontag's distinction, rather than "deprive something of meaning." But there is a reason that metaphors are what language is made of. They help guide our thoughts and make connections. That we *choose* our metaphors means we have the ability, the capacity, to choose metaphors of care and not war, of teaching and not destruction.

In this sense, "plague" remains an instructive word: "It is usually epidemics that are thought of as plagues," Sontag wrote: "And these mass incidences of illness are understood as inflicted, not just endured. . . . Diseases, insofar as they acquired meaning, were collective calamities, and judgments on a community."

A plague is a social phenomenon, not an individual illness. For a period—sometimes brief, sometimes a decade or more—in between a unique virus's appearance and its effective prevention and treatment, it really is plague-like: it strikes everyone who comes in contact with it, and it is the *community's* responsibility to prevent it from reaching those most vulnerable. But once the vaccine or therapy arrives and becomes widely available, the virus is again re-individualized: so-and-so has it but so-and-so does not; so-and-so is vaccinated and so-and-so is not—the difference being, usually, access to care or treatment.

We can carry this thinking beyond the plague months or years of COVID-19, and indeed beyond COVID-19 itself, if this virus does ever leave us. By the simple nature of having a body, all people are at risk of infection, or injury, or illness, or death, and moreover we are all connected, your illness to mine. The fact of all our bodies, vulnerable together, necessitates mutual care, including healthcare.

If we invert the metaphor of war into one of care, the question inverts itself as well. What must we do, for example, to care for our essential workers? What must we do to care for those most vulnerable in our society? There are layers in our answer to this question, and a time frame to consider. In the immediate future, labor's value must shift to reward those most at risk for their work that is allowing the rest of us to continue eating, to have hospitals waiting for us, and to have transit of any kind available to reach them.

So, too, must governments—state and federal—vastly increase their emergency relief efforts, no matter how "socialist" they seem. As illustrated in all countries that successfully contained COVID-19, people will only stay home if you pay them to stay home.

It's difficult to imagine, given the modern rhetoric of battle and the association of health with social class, but one of the oldest metaphors for Western medicine is one of care, not of war. Hippocrates wrote of "instruction in medicine" as though it were "the culture of the productions of the earth. For our natural disposition, is, as it were, the soil; the tenets of our teacher are, as it were, the seed; instruction in youth is like the planting of the seed in the ground at the proper season; the place where the instruction is communicated is like the food imparted to vegetables by the atmosphere; diligent study is like the cultivation of the fields; and it is time which imparts strength to all things and brings them to maturity." The germ, so to speak, of Western medicine is not invasion or colonization but *germination*—the cautious and patient nurturing of the body and mind so they can heal, and then thrive, among others.

Doctors and nurses aren't warriors; they are caregivers. They aren't heroes; they are workers. We owe it to them—when we think and talk about them this way—to stay home. They aren't dying "for our freedom." They are dying because we refuse to take tried and true precautions. Because they too have bodies, ones we have chosen not to take responsibility for, not to care for as well as we could. It is every American's responsibility to make sure their workplaces are as safe as possible.

Medicines aren't weapons; they are treatments necessary to live. At least in the best-case scenarios they are, as with HIV after 1996. But the HIV example, too, has its political dimension. When the medicine is a *weapon* in the war against HIV, we see who gets left behind, who gets sacrificed: largely Black queer people in the rural South, not to mention millions of people in nations overseas. War assumes the necessity of casualties, and often bases its moral legitimacy on these casualties—these actual deaths.

Care does not. Viewing medicines as treatment and care would initiate a reframing of our drug discovery process from beginning to end. Again, the HIV example is instructive. It took decades for scientists to discover the obvious: if everyone who has HIV in them is on (expensive) treatment, we can stop new infections from occurring altogether. Treatment *is* prevention. Treating HIV-positive people with the best care serves us all, HIV-negative *and* -positive. Mathematical modeling shows that we could simply treat our way to no new HIV infections. What stands in our way? Hospitals 2 hours away, no testing in rural areas, stigma, the cost of the care (and not just the drugs) itself.

Instead of the medications that would make the most profit by (for example) extending a patent for a condition where the generic is already just fine, we could finally begin researching novel medications—always much more resource intensive and therefore costly, therefore unprofitable—for diseases that are rare or primarily affecting people without sufficient medical resources. Instead of an "arms race" between profit-driven pharmaceutical companies, care imagines a cooperative effort to save and improve our lives rather than use and dispose of them. Care allows—no, demands—that we rethink systems that have been failing for decades, systems whose deadly consequences are too often made invisible. Care makes death invisible no longer.

This pandemic has shown exactly which jobs are truly essential to our survival. It is both shocking and not that many of these jobs are maligned, working-class professions. It is not at all shocking that these industrial, service, and healthcare jobs that keep us all afloat are so often held by people of color. If we were to "return to normal," where these positions remain underpaid and the "hero" status is demoted back to derision, that would not be care.

In the late 1970s, writer Audre Lorde was diagnosed with breast cancer, treatable then only with surgery and radiation. In the hospital before, during, and after this treatment, she kept a journal, which she published later as *The Cancer Journals*. "What is there possibly left for us to be afraid of," she wondered, "after we have dealt face to face with death and not embraced it? Once I accept the existence of dying as a life process, who can ever have power over me again?"

"The only answer to death is the heat and confusion of living," writes Lorde, "the only dependable warmth is the warmth of the blood. I can feel my own beating even now." Facing this moment, this global pandemic, this mass death, must give us fortitude, strength, and deep appreciation for our bodies and one another. We cannot accept going back to the death-making ways of capitalism, but must embrace the messy confusion of living as best we can. Care is how we do that. Care is insisting on the value of human life, every life, regardless of its position in a social hierarchy or the type of labor that person performs (excepting those, perhaps, whose profession or status is itself a form of harm, such as ICE officers and billionaires, but getting rid of a profession doesn't mean one can no longer care for the person once they're forced into retirement). The metaphors and realities of capitalism—war, imperial expansion, battle, competition, colonialism—are buried deep in our society, but they are not, so to speak, incurable. As Biss wrote, "The extent to which it is hard to imagine an ethos powerful enough to compete with capitalism, even if that ethos is based on the inherent value of human lives, is suggestive of how successfully capitalism has limited our imaginations."

Jose Esteban Muñoz wrote, "Capitalism, for instance, would have us think that it is a natural order, an inevitability, the way things would be." Ursula K. Le Guin framed it plainly in a speech: "We live in capitalism. Its power seems inescapable. So did the divine right of kings. Any human power can be resisted and changed by human beings. Resistance and change often begin in art, and very often in our art, the art of words."

Where Biss, Muñoz, and Le Guin all agree, can we really dissent? Our mission now is to exercise our imaginations outside the limitations of capitalism like a muscle, to begin to believe we can be the ones who can write a different future.

Everyone, every person, and not just writers, engages in the art of words. We all speak. We all write. We can all—not just professional writers—choose a language of care, a daily art making of connection, and love.

"We are those who are at risk with each other," wrote science studies scholar Donna Haraway. Facing COVID-19 and climate change, we are a global human whole likely to rise or fall together. We can see our connections clearly now. "We are symbiotic systems; we become-with relentlessly. There is no becoming, there is only becoming-with," Haraway writes. As an enmeshed, biopolitical ecosystem, we must *become-with* COVID-19, not fight against it. Vaccination, for example, allows our immune system to become-with, to be in conversation with, the SARS-CoV-2 spike protein without having been infected with a virus. Who has access to this conversation, now, becomes central to our rhetoric of care; until we protect every person on this planet with a vaccine, SARS-CoV-2 may well evolve to be better at evading my own protection, my own vaccination. We all become-with together.

And so it *is* possible to imagine alternatives based in care, in community; and the imagination begins—it ignites—at the level, with the flame, of metaphor.

If we continue the "fight" against COVID-19 as an all-out "war," we will continue to lose. We will continue to accept, not protect against, death at a massive scale. War, in relation to COVID-19, always described this death. It foretold and accepted—even dared—death, not our mobilization against it.

If COVID-19 is to remain for a year or more, as virologists predict, as it now already has, our plague, let it teach us that it's not our fault, that it can't be fought, that the sick are not weak and the dead did not fail, and that our leaders cannot rely—now and afterward—on the specter of violence to entrench us in our isolation. COVID-19 vaccines have shown themselves to be incredible biomedical tools; let's prove ourselves worthy of them. If by some grace or magic COVID-19 comes and goes, it will be integrated into our bodily and cultural memory; we must write it there through the framework of how we cared for each other, not simply how we fought, struggled, battled, suffered. Remember me. Remember this. Let it teach us, instead, how to survive a plague together and how to make sure the world will never be the same.

7

On Mentors

Some Other I before Me

Like most queer people, my parents are straight. I was raised in the country, in rural Washington State. We couldn't get cable, so my dad installed an antenna rotor on our roof to help with the salt-and-pepper static that clouded NBC, but this was before *Will & Grace.* All I knew about being gay is that it would get you beat up as a kid, and then you'd die from AIDS.

I was in college before I learned how gay men had sex. I'd been attracted to men by then—my first boy-crush was a lanky, tall boy who played basketball in the intramural league that I organized for my work-study job. Still, I had no idea what two men who liked one another did about it. I was sure they kissed. I'd watched short video clips of men jerking off, that I understood from my own practice. What more did they do? How did gay sex work? I couldn't picture it.

If I had no idea what sex between men consisted of, intimacy seemed even more opaque. How to be intimate, how to touch, without being called a fag, the worst thing at all to be called, especially if you worried that it might be true. Intimacy between men was then—like it sometimes seems even now to be—a foreign language spoken in a country where I was only a visitor, dependent on a dictionary even for banalities: ordering an orange juice for breakfast, a gin martini at a bar, and what exactly does it mean when a man grabs your upper arm as you walk, and is a smack on the ass just encouragement for a good catch on your intramural Frisbee team?

"What did you think gay men do?" a straight friend once asked me, astounded at my naivete. Two weeks after that conversation, home from college on spring break, I stuck two fingers inside myself and vowed that after college I would end up in a city where I'd be able, finally, to learn.

....

Alexander Chee, smirking, in black and white, stares at me from the cover of his new book of essays, *How to Write an Autobiographical Novel.* Inside, there he is again, a grid of him, smiling widely here, flirting there, making, in my favorite image, a kissy face. These photos were taken in a photo booth in Iowa, where Chee did his MFA, to send to his then-boyfriend. They're intimate, sexy, playful, a glimpse into his life and a relationship he had, and lost, decades ago.

Like me, Chee grew up before the Internet and in the middle of nowhere. Unlike me, he wasn't just nerdy and gay, he was also of Korean descent in a deeply white town. Chee's book begins in childhood, but not in his native Maine. We start abroad, in Mexico, where he finds in his foreignness a sense of belonging he never had at home. This theme—where we belong, and to whom—runs through all the essays in Chee's book, which act as a memoir in pieces rather than individual essays themselves.

Chee follows the thread of his life from Maine to college to San Francisco during the height of the early HIV/AIDS crisis. He met trauma early in life, with the death of his father, and then came of age in a world where gay men were dying of AIDS.

There is connective tissue between all his essays. Although the form changes, from a lyric memoir of a rose garden to a listicle on writing and finishing a novel, the essays return, one after another, to the book's central questions: the HIV/AIDS crisis, Chee's childhood abuse and decision to write through it as a fiction, trauma and healing, reading and writing, loving and breaking our patterns of false love, and making art in a world that undermines, even mocks, art's value.

Here's the strength, the brilliance, of Chee's book. He hints at these things, but then withholds detail until much later. We learn of his childhood abuse in an early chapter about his relationships in his 20s failing. Great writers, particularly ones attuned to plot, don't put a gun into the hands of a character unless that gun will be fired. We wait, and wait, and years and pages later, the trigger finger finally twitches just far enough. This—and Chee's trademark, precise but wandering sentences—pulls us rapidly through the chapters. But beyond its success as a narrative device, this is what trauma feels like: a loaded gun, always there, without lock or safety, ready, at all times, to go off, to break us, to explode the world we think we've safely built.

....

I kept my promise to myself. I moved to New York in 2006, after college and a year abroad, for a PhD program in biophysics. I was surprised, although I suppose I shouldn't have been, to find many gay scientists my age in my program. There weren't any out gay faculty members, though. It took some time in this new, big city, but I found many gay friends as well—all my own age—outside of science. In school, I had many mentors, all straight, all scientists, all older. My scientific mentors offered much wisdom, but their advice arrested short of the biology of one particular body: my own.

Queer people often find our kin later in life. We find mentors, those who have survived long enough to have accumulated some wisdom and who are willing to share it to make our own survival a bit easier.

Except when we don't.

I was born only a year before HIV was conclusively shown to cause AIDS. When I moved to New York in 2006, it was to a city where so many gay men five or ten years older than me had not survived a plague. I was 13 in 1996, the year protease inhibitors were added to the drug cocktails to treat AIDS. This is the year the HIV-positive started to live.

"I was always having to be what I was looking for in the world," Chee writes, "and wishing that the person I would become already existed—some other *I* before me." As queer scientists, my friends and I understood this all too well.

This sentiment can't be disconnected from Chee's race or his queerness or his ambition to be a writer. It also can't be taken out of the context, as he writes earlier in the book, of the HIV crisis. He arrived in San Francisco in 1989 and "had to overcome the false impression that no one like us had ever existed before, because the ones who might have greeted us when we arrived were already dead."

Chee writes: "We lacked models for bravery and were trying to invent them, as we likewise invented models for loving and for activism."

I arrived in New York in 2006 with much the same feeling, that the older queer men who might have welcomed me had gone a decade before. My friends and I felt the same way, that we had to invent a way to live because our lives were so different from anything we had been taught.

I didn't see answers for my questions in rom-coms. I'd read elegies, tragedies, but I hadn't read novels about gay people who'd made it, who'd survived. In queer work from, during, and before the plague years, I saw queer communities as almost *too* intergenerational: I was uncomfortable with some of the relationships that read like daddy/son, that seemed to introduce a young gay man into the ways of the gay world only through a sexual relationship with an elder. These relationships were problematic at the same time as they were easy to romanticize: at least someone was there to show them how to be, where to go, how to douche, what poppers are and how to use them, how public sex and relationships can go hand in hand.

I didn't see examples of queer lives lived in the mentors I had in science—while they were often open about their relationships, they looked a lot more like my parents' marriage than my own messy life, so full of false starts and huge losses. I didn't see answers. And so, it seemed to me and my friends that we had to make them on our own.

Maybe everyone feels this way, that the lives lived by their parents can't teach them much about how to live their own lives. Maybe everyone in their 20s is searching for a kind not-parent not-lover to help them into the world. Maybe, though, this is the curse and the blessing of queerness. Maybe each (gay) generation since World War II has had to reinvent itself outside of the nuclear families we largely come from and into our new queer possibility, a horizon we walk ever toward. We have had to deal with the loss of many of those we might have followed on our journey. I can't imagine the pain and loss of being in that generation. But I know, too, the loneliness of growing up gay in its wake.

....

Chee's book is a bit of a trick of false advertising. Its title suggests a *How to . . .* , and the book almost, kind of, delivers. The title suggests that the book is about writing, and it *is* about that, but not just that.

How to Write an Autobiographical Novel is a remembrance of a life lived and the country it was lived in. Chee remembers. He remembers the HIV era in a different way than I do. In an early essay in his collection we see him at an ACT UP (AIDS Coalition to Unleash Power) demonstration putting his hand on the side of an ambulance that's holding an injured friend, his hand a signal—to the police—not to harm him as well. A signal that he knew, and that we know, might simply be ignored.

In "After Peter," an elegy for a lost friend, Chee shares the weight of that time, how badly it could hurt to lose a friend, an acquaintance, an artistic and beautiful young man whose possibility was blotted out so early. "The men I wanted to follow into the future are dead," Chee writes. Losing a single friend is "a permanent loss of possibility, so that what is left is only ever better than nothing, but the loss is limitless," like "stars falling out of the sky and into the sea and gone."

Chee lived, and tells that story. I read "After Peter" for the first time on the plane, sobbing between two strangers, for this young man Peter and his friend Alex who, HIV-positive and HIV-negative, put their flesh on the line.

Memory and experience are the nucleation point of wisdom. They're not sufficient; so many people let the worst moments they live through turn them against themselves, make them hide from intimacy, hide from seeing themselves honestly. Life becomes a running away. Chee, according to his book, spent decades like that.

But, he tells his writing students, "Pain is information. . . . Pain has a story to tell you. But you have to listen to it." Chee's youthful mistake was believing that money, or success, or love, allowed us to control our suffering. "Money is not power over pain. Facing pain is."

It would be so easy for gay folks my age, in New York City, in the era of Truvada, to forget the pain and loss of the HIV crisis, a crisis that persists in so many places, still, in this country. It would be so nice to get two or three or more doses of Pfizer and go back to the club as if 2020 were an aberration. We forget our trauma at our own peril. Pain is information. The memory of what came before, and the willingness to acknowledge the pain of it, might be the first thing we ask for, that we need, from a mentor.

....

I first made older gay friends through my writing. In particular, it was my writing on growing up in the aftermath of the AIDS crisis, how the disease changed my relationship to my body, to pleasure. I was living in New York, I was already in my 30s. Gay writers who had survived that time were—they told me—just glad we were still talking about HIV. They were worried about being forgotten.

The first older man who slid into my Facebook inbox wanted what a lot of fake-ass mentors want. He was a poet—I should have known it would not go well—and invited me to coffee. He came all the way downtown to meet me in the middle of a snowstorm. We flirted in the easy way that gay men often do, and he hinted again and again at his closeness to so many famous writers, so many literary agents. It became clear that he'd read about a third of one of my essays. This was my first indication that he was more interested in me than in my work.

When he texted me to come uptown for Netflix and bad Chinese delivery, I couldn't even feign surprise. I lived with a boyfriend and our dog, Wesley and Winston, then, so I had an easy excuse. When his subtle offers to connect me to agents disappeared, alongside his ability to make eye contact at readings and events, I knew exactly what he had been after.

Many others, though, gave time, gave advice, gave love. And then there was Alex. I met Alex, in the flesh, at the Tin House Summer Writing Workshop. He was teaching, and queer folks I'd befriended were taking his class.

We became friends quickly, bonding like good homosexuals over rosé after most of the straight faculty and students had gone to bed. Alex had a way of arriving in a conversation, sitting and listening, cracking a few jokes, and then, in a moment of silence, dropping in a perfect nugget of wisdom, the kind that concludes the discussion of that particular topic, for good.

I met Alex at this writer's conference, and I met Tommy and Fran and Denne Michelle, too, friends who I would go on to collaborate with for years to come. We were there just over a week. That week, and the relationships it granted me, changed the course of my life. I can only see this in looking back, the way most things make themselves known. At the time, it was just rosé outside, laughing together, and feeling known and held from the moment someone introduces themselves, hiding behind a shy smile.

In his craft seminar at the conference, Alex talked slowly, deliberately, with many pauses. His audience leaned forward, leaned toward the language, jotted notes. "A novel," he said, "is not simply the story of a life."

A memoir, I wrote in my notes, isn't either.

Later that year, my partner moved out to transfer to a similar job abroad. One of the reasons he gave for leaving me was that he was worried that I would write about him, like I'd written about other exes, and he was tired of living that way. I messaged Alex, asking about the relationship between his writing and his love life. From reading his essays, including the ones that are now collected in his book, I knew he'd written about former loves and lovers.

"I deal with things," I said, "by writing through them."

"It sounds like you have a writing process that is hard to survive, and it makes me concerned," Alex wrote to me. He instituted long ago, he told me, a three-year rule, something he learned from Annie Dillard. He doesn't write about anything until three years have passed, and maybe even more when a man is involved.

"Living people," he texted me, "live uncomfortably in prose."

But, he said, people who have a problem with your writing will, have a problem with your writing. He says the same in his book, "Anyone who saw themselves in your characters will mostly see themselves, even if they were not described."

Alex checked in on me every couple of weeks for months after that. He offered words, a drink, dinner. He offered friendship.

Like Chee, I'd spent much of my life running away from pain and into the arms of willing men. "There was always a new man, another will-o'-wisp of desire," Chee wrote, describing his life as he moved from place to place, from lover to lover, never stopping to confront himself.

I asked Alex when he met Dustin, his partner. The answer reassured me—I was 34 at the time. They'd met when Alex was in his 40s. What reassured me even more was the simple fact that Alex and other gay friends of mine, now in their 50s and 60s, had lived through what I was living through. The simple fact that they'd survived was wisdom enough, as I wasn't sure that I would. "You can lose more than you ever thought," Alex wrote, "and still grow back, stronger than anyone imagined."

....

I'm a writer, and Chee's book is ostensibly a *How to.* I don't write much fiction, except the one terrible, fragmented novel I used, for two years, to query literary agents. It won't ever see the light of day, and that's fine. The first novel you publish, Alex Chee tells us in *How to Write*, is "almost never the first novel [you] wrote."

The first book I published was not the first book I finished.

I'm a writer, but I never studied writing outside of a few workshops. As a college student I studied literature, yes, but in French, and even then my main focus was biology. I learned to write because I cared about writing, and I cared about reading, and I wanted my writing to be good. I got lots of help with craft from friends—my own age—who cared about writing like I did. I got help from them, too, for editorial contacts, connections to agents and editors.

I was learning my craft by doing and by figuring out what worked. I needed mentorship to figure out how writing fit into my life and how to write about myself without undermining the possibility of where my life would take me. I worried that I was so open about queer sex, love, and loss that I might never get a job. That I might never keep a man.

What I needed, I thought, was guidance. In part, I just needed reassurance. "I was someone who didn't know how to find the path he was on," Alex writes, "the one under his feet. This, it seems to me, is why we have teachers."

Books can be teachers too. I needed to know that I could write about my own pain without being doomed to repeat it. I needed to know that I could write about suffering without seeking more suffering out. I needed to know that writing could free me of my past, not trap me in it. *How to Write an Autobiographical Novel* is an answer. Writing isn't healing, Alex's work tells me, but it doesn't have to stand in its way, either.

....

I imagine the gift this book might have been if I had met it in my childhood. The list of books I wish I'd known is long. Words by Anne Carson. James Baldwin. Maggie Nelson. Susan Sontag and Audre Lorde. The image of this bookshelf—the books I needed and didn't know already existed—is bittersweet. The sweetness is how much the books would have offered me. The bitterness is how much I needed it, and how late it came.

I have this bookshelf now, full of these authors and their work. What luck I have to be alive and to know them all.

One thing Alex told me as a friend, over a drink, is that gay writers need to allow themselves to be self-indulgent, to get over our fear of being read as dramatic, as camp. Here goes.

Scene: I'm me, it's 2018, Donald Trump is president and the world is a trash heap. I am a teaching professor of biology at NYU. I am a reader, but I never started writing. I never made any of the friends I made because we all try to do this impossible-seeming thing, taking a blank document and making it sing. I never met mentors by writing about an HIV crisis that some older writers had survived. I don't know Alex, I don't know John or Tommy or David or Randall or Darnell. I don't know any of these people. But I know their work. I'm sitting in my office on the weekend reading, because I love reading. The book is fire-engine red, its author stares out at me. In the pages I find the voice of a man who lived a life kind of like mine, loving and losing men, suffering and healing from suffering. In his words, I find memories of the AIDS crisis, the years I watched on TV. I lean forward, trying to hear the silences between the lines. On the pages, I find guidance, I find hardship and truth. I find love. I find a model for living a creative and true life in my 30s, my 40s, my 50s. This book is not flesh, it's not bone, it doesn't breathe, but it is friendship, and care, and love. It's 2018, and all of these things are in short supply. I am 35 and crying at the fact that I never had this mentor, and the fact that I've finally found him in these pages.

Scene: It's 2018, and I am a child, 12 or 13 or 14. It's me, but not me now. When I was 14, *Dancer at the Dance* was not available at the library in my logging town, and I wouldn't have known to look for it anyway. But now it's 2018, and this book *How to Write an Autobiographical Novel* is big enough to be written about on *Buzzfeed* and in *Teen Vogue* and I see it. I am wondering about my sexuality, and I want to be a writer—maybe—someday. I ask my mom to buy the book, and she does, thinking it's a simple *How to.* In its pages, I see myself reflected and refracted, I see a future that's mean and big and true. I cry, and I know that life will not be easy. It's better to know. I wish I had known; it would have made my difficult life easier to love. Zoom out and look at the book, fire-engine red, there in my hands, as I sit on a school bus. No one knows it's a gay book, and so I'm safe. I see, on that bus ride home, the world grow so much larger, my own future suddenly inevitable, and possible, and grand.

Scene: It's 2018, and I have Alex and his work, and I well up with gratitude for the gifts that writing has given me. The flesh? Well, I get to love on that too, and on good days I wake up beaming with gratitude for the gift of his gay friendship, his queer mentorship. I know it's precious, because I remember life without it. It took me three decades and more to find. His book now sits on my desk, next to the molecular biology quizzes I should be grading. The pages of his book are bent and broken, my pen has dug into the flesh of its paper. My eyes, open now on this page, imagining the future, knowing that it's coming. Now there's no going back. The world needs changing. At a recent talkback about his book, Alex told me that we're trying to change the world. We're crazy enough to believe that writing can do its part. Now there's only giving forward.

8

On Whiteness

The Sovereign Freedom to Harm

When the news told Americans that COVID-19 was still only in China, and when we saw the first glimpses of the data on who got the virus and who was mostly likely to die (older people, mostly), some in the media argued that COVID-19 might be a great equalizer. *We* had to work together because the virus was a threat to *us all,* they said. Susan Rice, for example wrote a March 13 *New York Times* op-ed titled "Viruses Are Equal Opportunity Killers." This virus, we thought, would be as likely to kill anyone's grandma as anyone else's. This was meant to mobilize us to action, to protection; we were all, all of us, at equal risk.

The "we" I mean here is the liberal humanist American popular imagination, the collective we of our nation that speaks in media and pundits, but also in small conversations that bubble up and down from there. Most Americans probably thought the virus wouldn't impact them at all; many who did see it as a threat ignored all evidence that viruses, like all diseases, affect the most vulnerable among us the most. Those of us in the COVID-19 Working Group—the activist organization I'd been a founding member of—knew by early March that homeless New Yorkers and people in jails and prisons would be disproportionally affected; we knew that "essential workers" who couldn't work from home would be putting their health on the line for us all. We knew, too, that these individuals were more likely not to be white, not to be wealthy, not to be a part of the media, academia, or the government.

Less than a month after the article by Rice, the *New York Times* published, on April 8, the headline, "Virus Is Twice as Deadly for Black and Latino People Than Whites in N.Y.C." Work by one of our founding organizations—Black Health—was essential for getting the city to release COVID-19 data by race and not just by zip code.

By mid-April—once we knew who was most likely to get sick and die—our federal government was pushing us to fully open back up, even as the cases of COVID-19 surged in too many places.

The first pivot out of the shutdown came as we were realizing just who was most at risk. Who are we willing to let die for American normalcy? American as apple pie.

This is not because racism is a virus. It is not. Viruses are part of the natural world. Racism is always human-made.

It is true that viruses and bacteria don't discriminate by race. They don't *see* race because they're biological entities and race is distinctly not. Decades of work by sociologists and biologists have shown that race is a social category, not a genetic background. It is impossible to tell if someone is white or Black even by sequencing all of their DNA, because the history of this country means that some white people may well have Black ancestors who weren't ever known about or whose history has been erased. Ask that White Supremacist who breaks down on the *Trisha Goddard Show* when he finds out that his DNA shows 14 percent sub-Saharan African genetic ancestry. It's not "statistical noise" as he claims, but those DNA sequences of probable African origin doesn't mean he isn't white. He's white because he calls himself white, and his family did, too. Many Black Americans also have significant white ancestry in their lineages, the legacy of rape during enslavement and after, and the one-drop rule, among other American horrors.

Blackness, Brownness, and Whiteness are not biological, but they are biomedical. The difference is medicine, another human invention, another social and sociological construct in addition to a specific set of places (offices and hospitals) and interventions (checkups, drugs, and surgeries).

Doctors take the pain of Black patients less seriously and are less likely to treat it. Black women are more likely to die in or after childbirth. Rates of many diseases, including cancer, heart disease, diabetes, and autoimmune diseases, are higher in Black and Brown people than in white people, and they often have higher mortality rates, too. Hospitals, in 2020, in Black and Brown neighborhoods weren't as good as the ones where white people live, and so the quality of care that Black and Brown patients received when they got sick with COVID-19 was lower.

This has everything to do with race and class and geography. Health insurance is needed to manage our bodies in the way society tells us we should. We micromanage our health, eating right, seeing the doctor, exercising, taking supplements, taking meds. All of this costs time and it costs money, but good people, healthy people, do it, right? Eula Biss's notion of good-health-as-identity is indeed raced; it's simply another way to blame Black people for their own deaths even as *we* have built a system that creates those deaths.

An extensive literature by scholars like Harriet Washington and Alondra Nelson and Dorothy Roberts tells us that it has never not been like this. American medicine was built using Black people as research subjects and then built with barriers between Black Americans and the medicines their bodies were used to make once the medicines were deemed safe enough for *us*. The American *us* has always been white.

J. Marion Sims developed antiseptic surgical procedures and silver sutures to decrease the risk of infection. Sims is considered the father of American gynecology, but he worked on enslaved women who didn't need his surgeries, who he sometimes even bought to be subjects in his research. Informed consent, any consent really, doesn't even enter the question. What can property refuse anyway?

Indigenous peoples on this land were systemically wiped out in part by smallpox brought purposefully on them by the colonizing people from Europe.

HeLa cells were taken, without permission, from a dying Henrietta Lacks, and her cells live forever. They've been used to test drugs and vaccines for years, all while her family lived in poverty. They've been used to test HIV drugs, to test COVID-19 drugs. I've grown her cells, think about that, *her cells*, myself, in a lab, in a petri dish.

In Tuskegee, Black men with syphilis were not treated for the illness for decades so that researchers could determine the "natural" course of the illness. Many died, even after the antibiotics that could readily treat the infection were widely available. Researchers lied to the study participants, acting as though they were giving free healthcare, free yearly checkups, in this area where many people went without. The experiment ended only in 1972; this isn't ancient history. Tuskegee might be the most widely known example of American biomedical racism, along with Henrietta Lacks, thanks to the tireless work of Black scholars and storytellers. But when told in the mouths of white people, the assumption is usually that these were horrific outliers in the norm of science-for-the-good-of-humankind. Rockefeller University, where I did my PhD, calls its mission "Science for the benefit of humanity." That makes one wonder how the university sees humanity; when I attended, of its 100-plus faculty, none were from underrepresented groups as defined by the NIH. It is so easy to imagine the scientific method as inherently good. Science is but a tool; it's just as easily used to harm. The examples of harm in science are not the outliers; science has always been embedded in White Supremacy.

The birth control pill was tested, at incredibly high doses, on Brown women in Puerto Rico without any informed consent, leading to lifelong issues with fertility. Growing up, I always wanted to live in the uninhibited, protest-filled 1960s. I hadn't yet learned whose bodies were sacrificed for all that white American freedom.

Early human geneticists trafficked in eugenics, trying to understand the deficiencies of any and all nonwhite people, trying to improve the human "race" by selective breeding, as detailed in *In the Name of Eugenics*, by Daniel Kevles. In the struggle to make birth control accessible to women in America, Margaret Sanger (the founder of Planned Parenthood) allied herself with anti-Black eugenicists.

Harriet Washington's *Medical Apartheid* details a researcher at Columbia University, in the 1990s, giving the mood altering weight loss drug Fen-Phen (Fenfluramine) to 12-year-old boys—children—whose older brothers had been involved in the juvenile justice system to test her hypothesis that they had inherited aggression disorders.

HIV taught us that viruses were one of the many vectors that Whiteness can use to kill or harm. And that, once you've begun viral surveillance (or worse, viral criminalization), it can be hard to undo and easy to abuse.

In 2013, Michael Johnson was arrested in Missouri for "recklessly infecting another with HIV" and "attempting to recklessly infect another with HIV," both felonies in that state. His case was covered extensively for Buzzfeed by journalist, scholar, and writer Steven Thrasher. Whether Johnson had indeed infected anyone with HIV was unclear; although molecular tests exist to trace such an infection, those tests were never done on Johnson and his sex partners, including the only one that tested positive after having sex with Johnson.

Michael Johnson is Black and—at the time—was a college wrestler at a primarily white institution. As detailed by Thrasher, his white sex partners fetishized Johnson and then reacted with predictable violence when—after consensual unprotected sex or oral sex—they found out that Johnson was HIV-positive. The risk, indeed the guilt, of HIV transmission was placed solely on Johnson's body and person, even though two partners in each hookup actively consented to the risk of sex without a condom or PrEP.

HIV criminalization laws are common and widely known to be ineffective. They don't reduce the rates of HIV transmission. They place the entire burden of consensual sex onto one individual; in some places, even low- or no-risk sexual contact without disclosing an HIV-positive status is still illegal. Dating from the first HIV plague and HIV panic years, these criminal statutes are not based on science, they do not currently serve the HIV-negative or -positive communities, and yet they're still on the books in 35 states as of 2021.

Johnson, after a trial steeped in homophobia and racism, as reported by Thrasher, who attended every minute, was found guilty. He was sentenced to 30 years in prison; he served 5 years before being released on appeal.

Once a disease is criminalized, the fear of testing positive can discourage testing. Why get tested if a positive test turns you into a literal biohazard, every fuck a potential felony?

An essay published by scientist and writer Hala Iqbal demonstrates the lengths to which the US national security infrastructure will go to collect data, no matter who it harms. In the 2010s, the CIA set up a fake hepatitis vaccination site in order to collect DNA samples from individuals suspected of living in Osama bin Laden's compound in Pakistan. News of this fraudulent vaccine drive confirmed the worst fears of those in that country, leading to increased vaccine skepticism and outright bans on childhood vaccination. To protect people in Pakistan from COVID-19, we'd need to get to herd immunity by vaccination; the US government—my government—might have already made that impossible.

Black and Brown people also have a long history of being under government surveillance, including biomedical surveillance, here in the United States. The infamous Tuskegee experiment followed a large number of Black men in the rural South for decades, not only refusing them treatment for a curable bacterial infection—syphilis—but also taking their samples again and again over time. The bacterium that causes syphilis was at the time almost impossible to culture in a lab; the government used tissue from these people—human men turned into effective bio-incubators—to make enough of the bacterium to develop diagnostic tests against it.

Thanks to the Patriot Act, law enforcement, including the FBI, the CIA, and ICE, may have access to federal, state, or private health information. From the 1950s through the 1970s, the CIA, through the COINTELPRO program, infiltrated, surveilled, and quite literally murdered Black activists from the civil rights movement, including members of the Black Panthers. Biomedical racism was so blatant, at the time, that the Black Panthers invested in free health clinics and worked with allied scientists to develop diagnostic tests for sickle cell anemia—a disease so much more common in Black Americans that it was mostly ignored by science. The American surveillance state did not end with COINTELPRO. Muslim Americans in New York were surveilled and their organizations infiltrated for years after 2001, mostly by the NYPD. The NSA collected metadata and individual phone records for countless Americans for years.

This history makes so many Americans justifiably wary of giving the government any private information.

This is the world we live in. This is the world we made.

In 2020, another pandemic. Medicine sees race, and the ramifications are deadly. Look to the last catastrophic pandemic, HIV.

Black and queer people, and Black queer people in particular, bore (and continue to bear) the worst of that plague.

Of course COVID-19 would kill Black and Brown people more. We have to eat, and who's going to deliver the food we need when we get home from (log off after) a long day on Zoom? *We* weren't ever locked down; *we* stayed home, and passed our risk along to other people, poorer and more likely to be Black or Brown. *We*—white Americans and the infrastructure we've built—have always allowed and perpetrated these deaths, especially when they happen well out of sight.

This is nothing new. "Crisis creates a sense of ahistoricity," I heard Mabel Wilson, professor of African American and African diasporic studies, say during a panel discussion on Black Lives Matter and COVID-19.

This is nothing new. And yet, it still horrifies: COVID-19 can kill everyone equally, but it kills Black and Latinx and Indigenous people disproportionally, so it's not worth closing restaurants, it's not worth closing schools, it's not worth distancing or wearing masks, and football? With its majority Black and Brown workers? With its massive teams and traveling schedule? *We* must be entertained! Whiteness is the freedom to do harm. Play ball.

....

To paraphrase Toni Morrison, white people have a very serious problem. To paraphrase James Baldwin, if your humanity requires the subjugation of other humans, what does that say about how you see your true worth, your own humanity? To again paraphrase Toni Morrison, if you take away a white person's race, do they still feel strong, feel smart, feel tall? If you only feel tall because someone else is on their knees, you have a very serious problem.

Capitalism atomizes and dehumanizes us all. We become not people, but workers, who aren't owed life but who owe something to society. We trade our labor for the ability to eat, to have shelter, to be living.

To oversimplify, capitalism is an economic system where the vast majority of workers trade their labor for a wage. Privately owned companies use this labor to make a product or provide a service, and the company maximizes profit for the owners and shareholders. The business owners are the bourgeoisie, the workers are the proletariat, and a governing class includes voices from both the bourgeoisie and (mostly) a monied, historical aristocracy.

All of this takes place in a magical place called The Market, where the Invisible Hand of how much consumers want to buy something (demand) and how much it costs and how hard it is to make (supply) sets everything's price. The needs of the workers (wages) and the owners (profits) are at odds; pay the workers more, and profit will decrease unless you increase the cost to consumers in the end.

This is how we are atomized. We aren't whole needy humans. We are our labor and our consumption. The more we work—and the more we consume—the greater our value to the class of owners and aristocrats who generally have the power to set national policy. And anything that drives one class of working people to despise another may help pit one against the other rather than against the ownership class that pays us too little. And the household labor of family rearing, mostly done by women, is invisible and goes unremunerated. And deviance from family and social norms, the desire to do things other than "productive" work, is punished. And anyone who has land that might be "productive" should give up that land to the capitalist classes. As Elissa Washuta writes in *White Magic*, considering Washington State, the ship canal, and the river dams that forever changed its waterways, we (white people) must dominate even the land itself. "Boundaries were set," she wrote, "between water and land because impermanence and change made for poor real estate. Nature wasn't good enough for settlers; it demanded transformation."

Even the land itself is not safe from the needs of Whiteness to exert power over it, especially if there is a neat profit to be made—and isn't there always, in the eyes of Whiteness, a neat profit to be made?

Capitalism may well not be the cause of all the -isms and -phobias, but it certainly is their common denominator.

Whiteness, I think, is a hundreds-year-old practice that attempts to reclaim our full humanity by interpersonal and systemic violence, through which we imagine we have agency, control. In *The History of Sexuality* and *Discipline and Punish*, Michel Foucault notes that the right to physically punish, to torture and even to kill, once belonged to a sovereign, a king. The right to harm another person was synonymous with power.

In these books and others, Foucault traces the history of this power once the right to take life often becomes a less direct mandate of government, of the state. He writes of schools, of biomedicine, of hospitals, of prisons as spaces that hold some of this power over life and death; these are the very same institutions we often speak of when we identify how systemic racism does its life-ending harm. Foucault doesn't sufficiently make this connection, to all of our detriment.

Luckily, there is also historian Achille Mbembe, who directly connects Foucault's idea of the sovereign and of "bio-power" (the right to give or withhold life) to race and racism. "The politics of race," he writes, "is ultimately linked to a politics of death. . . . In the economy of biopower, the function of racism is to regulate the distribution of death and to make possible the state's murderous functions." He ends the essay by arguing that slavery, racism, and colonialism create a "death-world" wherein the state (and other actors) work to maximally destroy as many (marginalized) people as possible; living in such a state creates, he argues, the status of the "living dead."

Luckily, there is also historian Cedrick Robinson. In *Black Marxism*, Robinson traces the origins of our modern notions of race to the emergence of capitalism from feudalism, which itself coincides with and depends on the Atlantic slave trade and colonization in Africa, the Americas, and Asia. Modern conceptions of race (including Whiteness and Blackness) come about alongside, and because of, capitalism's emergence from feudalism. It's not that feudalism was great: it was a brutal use of labor often with little or no compensation. Its *direct* brutality, Foucault argues, was replaced in capitalism with a more indirect, diffuse, but in many ways no less brutal system. The direct power of the sovereign over life and death was exchanged for indirect control through various institutions, including hospitals, banks, schools, and prisons. The decisions for the price point of a necessary drug isn't made by a king, nor is the decision to not cover abortion as healthcare, but both of these choices can easily put many people to death.

Robinson also delineates that race in the European context predates the arrival of European notions of Blackness. Before Black workers and enslaved Black people, there were Slavs and the Irish, who Robinson argues were racialized and exploited minorities. And who had power? The same people who colonized the Americas, the same aristocracy and burgeoning bourgeoisie that profited off this labor, from farming to trade to manufacturing, even in its infancy.

White people had power. White people were The Sovereign. They had the freedom to harm, to kill. But, over the hundreds of years since, Whiteness has mutated, incorporated more once-ethnic identities into its cult of exploitation, of punishment, of freedom to harm.

If the Slavs and Irish were racialized categories in feudal

Europe, as Robinson argues, it is obvious that Whiteness is historically mutable. People of Slavic and Irish descent (hello) are quite obviously white now, in twenty-first-century America. This shift began, Robinson argues, as the transatlantic slave trade remade the world's geography and markets and as the wealth it created made new classes possible in Europe. The economic classes consisted no longer of aristocrats and farmers; now we had traders and lawyers and doctors. Now we had—in England—landed gentry. All this new money! All these new classes! All funded by exploiting stolen land and not paying for stolen labor. The bourgeoise was born. Capitalism was birthing itself, slowly. Overwhelmingly, worldwide, the bourgeoisie was white, a new class and racial possibility. The wealth that funded industrialization, that cemented capitalism, invited yet new classes of people to move to urban centers and become urban workers? Once again, Robinson reminds us, that wealth was created by colonialism and slavery.

And here we are today, in a world that looks nothing like Mansfield Park, sure, but in a world that was formed by that world. Here we are; I am Irish American, and I am white. In the context of the United States in the twentieth century, Whiteness rapidly expands its meaning. Ethnic white people—German and Irish and Greek Americans—buy into Whiteness specifically by becoming as anti-Black and anti-Indigenous racist as the WASPS (white Anglo-Saxon Protestants), those who run the banks and the government. In *How the Irish Became White*, by Noel Ignatiev, and *Working Toward Whiteness*, by David Roediger, we see two authors grappling with this immense change in the structuring of race in our country. The thing that underlies both transitions—Irish Americans losing their hyphenate all days except Saint Patrick's Day and the creation of a panethnic working-class white identity—is the engagement with institutions that excluded Black people: unions, New Deal jobs, and home ownership.

These white ethnic groups finally perform and finalize their integration through anti-Black racism, a process that started under slavery and continued beyond it. In her novel *Libertie*, Kaitlyn Greenidge describes the 1863 race riot in which ethnic whites (then very much excluded from Whiteness) descended upon Black families, businesses, and abolitionists to "protest" the Civil War draft.

W. E. B. Du Bois explained it plainly in his essential work *Black Reconstruction* (also heavily cited by Robinson): "The system of slavery demanded a special police force and such a force was made possible and unusually effective by the presence of the poor whites . . . [slavery] gave him work and some authority as overseer, slave driver, and member of the patrol system. But above and beyond this, it fed his vanity because it associated him with the masters." From slavery, as explained by W. E. B. Du Bois, through the twentieth century, when poor whites fully integrated into American Whiteness, anti-Blackness gave white people power *over*. That's what we bought into. Not better working conditions (although they may be better than those of Black people), not better housing (just better than Black people's housing), not better jobs (just better than . . .), not better schools (just better than . . .), and always pull up the ladder to those—especially Black people—left behind.

Yes, of course, this is a way to divide and conquer, it's a successful method of oppression, even of white people. But yes, this is the path we chose. Between solidarity between workers to push for a better world and a racial hierarchy that accepts the deadly status quo for all races, white people chose. We didn't choose freedom. We chose the original agency, the feudalist tendency: the freedom to harm.

Whiteness is inextricable from the violence of our forebears, the genocidal displacement and death of Indigenous Americans, the transatlantic slave trade. *But my forebears didn't do any of that! They came to America poor, and look at us now!* Whether our individual forebears owned enslaved people or fought for the Union in the Civil War or just arrived a generation ago, American Whiteness is ours. As Du Bois reminds us, anti-Black racism was maintained by us. We might not have asked for it, but it's impossible to be a white person in America and not to own this history and to benefit from the power of Whiteness.

I grew up around a toxic Whiteness so saturated with toxic masculinity that was violent toward even me: a femme-of-center, bookish little white boy. Sometimes the violence was physical (two fists in my chest and the back of my head against a metal locker) and sometimes it was simply policing who I could be ("Why do you sit with your legs crossed, you look like a giiiiirl"). By and large, people were all poor. This always confused me, as a kid: *This* is what white privilege looks like? People back home were on food stamps and drove beater cars and made never-enough-money.

The white people I grew up around performed their race in F-150s and country music blasted on CB radios. I had to get out of there; there was no possibility for the human I was back home. I got to college and expected to be saved. But no, Whiteness there wasn't much kinder to me, a femme-of-center kid who barely had enough money to be there to begin with. The boys at college performed their race with vintage Saabs, Miller High Life, 90s rap, and date rape. I was an 18-year-old virgin with a limp wrist who couldn't afford a car. There was no space for me there, either. Here, too, my femininity made me suspect, other. I was shocked—I imagined rich, educated white people weren't like this.

But every time I walked into a room, people assumed that I was just like all the other white kids, which is to say from a prep school out East or a magnet public school in Iowa. In a sense, I resented this. I performed my race and class so that it would be seen: I wore a trucker hat around that said "Beer! It's not just for breakfast anymore" although I rarely drank. I wore deeply uncool cargo shorts and flip-flops and I said "ain't" and used double negatives, like everyone I grew up around. I resented the apparent invisibility of my class background. I wanted people to know how different my childhood was than what they might be expecting when they saw my face and body.

There was no healthy way I could have been white. Assimilation into Saab-Whiteness was literally impossible; I couldn't figure out the rules of the game of their social structures, and so I could never fit in. I didn't know enough about myself to perform my own race in ways that made sense, that didn't feel appropriative even of the people I grew up around because, to be honest, back home, I'd made fun of the kids who said "ain't," needing to feel better than them to survive the fist in my chest that sent my head into metal.

What a cost Whiteness is, even to us white people. This isn't unique to Whiteness, of course. Patriarchy harms men. As a boy, I committed the worst sin a boy can against white manhood: I cried. I cried all the time, ugly cried with snot dripping down my nose. I mostly didn't cry when I was sad, but when I was frustrated or let down or scared. I felt too much; little boys don't *feel*. I am ostensibly a man, and the way my every action as a child (my limp wrist, my love of baking) was policed did me lifelong harm.

In his posthumous book, *The Sense of Brown*, José Esteban Muñoz writes, "It is not so much that Latina/o affect in performance is so excessive, but that the affect performance of normative whiteness is minimalist to the point of emotional impoverishment." Later in the same essay he writes, "At this point in history, it seems especially important to position whiteness as lack." How I needed to hear this growing up. Cry, baby Joey, cry; the world is hard and it will only get harder!

If white patriarchy had allowed me, as a young boy, to feel, I would have grown up with so much more possibility. I couldn't not feel; I was just punished for it, mostly by other white boys and men.

Their race, and their performance of it, gave them permission—the agency—to do harm to me as someone who couldn't help but transgress the possibilities of normative white masculinity.

I want to burn it all down. Inside my body, writing this, I burn with rage.

In the post–World War II era, when so many ethnic groups, including my own, had new access to Whiteness, companies invented a version they could own. They could buy a quarter acre and a house in one of five floor plans. They could own a car and commute to their managerial position at the factory. They could move to those havens of emotional repression: the American suburbs. Histories of Robert Moses cutting freeways through the Bronx so that white people could commute to midtown tell the story. Joan Didion's California essays from the 1980s and 1990s tell it, too, culminating on my favorite essay in her large oeuvre: "Trouble in Lakewood."

Of course, newly white people were granted a Whiteness they could purchase, while others—white bankers and white builders and white real estate agents—made a neat profit. And I understand why some may aspire to Whiteness, especially given how precarious life in capitalism is. The lifestyle of the rich and famous can be so alluring that it's easy to ignore where their wealth came from and the moral and spiritual bankruptcy that it carries. Let's read Gatsby together.

Whiteness harms white people; as long as we're committed to White Supremacy, we will never be free of the economic violence of capitalism. As long as we view our American lives as freedom to harm, we will remain spiritually bankrupt. As long as we trade our relative American comfort for WASP cultural norms—don't cry in public! don't talk so loud!—our lives will be traumatically underlived. Whiteness and racism and patriarchy and homophobia and transphobia and the foundational force under all of them—

capitalism—are extinguishing our ability to continue living on this planet. When I say our, I mean our species. The capitalist class may be looking to Mars or New Zealand as an escape. I know there won't be enough space on Mars for me, and anyway, I wouldn't much fancy the trip. I'd rather go down with the ship. Earth lives, or we all die.

I know my life would be dramatically different, every day, in too many ways to count, if I lived in a nonwhite body. I cannot imagine the small and large traumas, the effects on my work, my body, my soul. Empathy can only go so far; we can listen and read and write and organize together. But when it comes to race, white people are always going to be students of the world because we cannot experience racism. Here, I'm defining racism not simply as bigotry—disliking someone for a characteristic of their identity—but as a structural force. Racism is something that systems of power have embedded, from Jim Crow to the Chinese Exclusion Act to Red Lining to contemporary school districts being funded by local taxes. Empathy with marginalized peoples is not sufficient to undo systemic harm; solidarity, in words and actions, is the only hope left.

In Tarot, one of my favorite cards is Death. I have a beautiful watercolor representation of this card—randomly sent to me with a deck—above my writing desk, next to a David Wojnarowicz photograph of his friend Peter Hujar's dead body. Only one watercolor print was sent with each deck. Mine was Death; I view this card as the one I pulled when I received this deck, and only use this particular deck when I need to break clean, rupture, and I'm afraid of moving forward or don't know how. The Tarot deck is cyclical: It's a story, and the story never ends. Death, then, is the end of one story, but always the beginning of another. Death solidifies the knowledge one has gained. Death comes from the wisdom one gains at the end of a journey.

To paraphrase Toni Morrison, white people have a very serious problem. She goes on to tell Charlie Rose, who is visibly uncomfortable, "And *they* should start thinking about what *they* can do about it. Take me out of it."

"Then give white people some free advice," Rose tells her.

"It's all in my books," she responds with a smile.

Take me out of it. What does Whiteness mean and look like in America without Blackness, Brownness, and Indigeneity? Who does it serve? It serves money more than it does people. It is a way to be inhumane and so—as Morrison puts it—"morally inferior." I think we know this to be true. We can choose, any day, to stop hurting ourselves and others. We can choose solidarity over the freedom to harm, and free ourselves of the (traumatic) confines of Whiteness in the process. I'm not optimistic, but I hope that I will live to see that day.

Death is the end of a cycle, the solidification of the knowledge gained from it. Whiteness must die. I would like to help kill it, if I can.

....

In July of 2020, Patrick and I were working on an essay about war, language, and disease that became Chapter 6 of this book. As nerds, between us, we read or reread a handful of books and essays, including Sontag and *On Immunity*, Biss's work on illness, identity, and history. I hadn't read *On Immunity* since it came out. Biss's notion of health-status-as-identity was central to our argument. Her writing on this topic is clear, and we cited it several times in that chapter. And yet I made a dumb Facebook post in which I criticized her approach to race, claimed she got a lot of the science wrong, and wondered why anyone liked the book at all.

My post was mean-spirited. I know why people like the book. I liked the writing about identity. That's why we used it. The sentences are beautiful. But rereading Biss's book shocked me. I was frustrated, probably first with myself. I'd read it when it first came out, and I don't remember this anger. Maybe the world has changed since *On Immunity* was published. Maybe I have changed, too. I think I was misdirecting my anger at myself for not seeing and speaking up sooner onto Biss, a classic projection meant to not feel bad about myself.

It's white writers' job, I think, to talk to one another about how we use race in our work, whether we interrogate our Whiteness. I think Biss wants to do well; we have friends in common whom I adore. But I can't use her own writing in my own, as Patrick and I did earlier, without exploring the underbelly of her work, the Whiteness that sutures it together, and the anti-Blackness that is inextricable from the Whiteness she doesn't sufficiently explore. I need to use her book in my own for its thinking on identity; I have to discuss its Whiteness because I don't think she does it well enough.

My problem with Biss's work is that she uses race, and American Blackness in particular, in deeply fraught and uncomfortable ways without ever explicitly laying claim to her Whiteness and class, the power they grant her, and her relationship to Blackness.

"Vaccination, like slavery, raises some pressing questions about one's right to one's own body." She writes this in the context of a discussion of vaccination and race in America, and about how nineteenth-century vaccine resisters used slavery as a metaphor for mandatory vaccination and invoked the rhetoric of abolition to support their own anti-vaccine position even though they had no interest in abolition itself. But even so, to many readers, nothing is like slavery, and certainly not vaccination. Biss does consider the fact that an enslaved African in America, Onesimus, brought the first knowledge of vaccination to our country. She talks about how infectious diseases have—for hundreds of years—disproportionally affected Black and Indigenous Americans.

But Biss never explicitly names her own Whiteness or the Whiteness of her North Chicago suburb community of mothers hand wringing about how, if, and when to vaccinate their children. Why should she have to do this? Everyone knows that Biss is white. But all people are racialized, even white people, and it is important—even essential, I think—for white writers to start writing explicitly about our race, about Whiteness. For Biss, implying her race and class, and that of her friends, is enough; the critique of whiteness is inherent in her writing. For me, this is insufficient; Whiteness should be named explicitly, as other racial categories are. Implied critique may be subtle and may invite in more readers willing to engage without being defensive. I also believe it is insufficient for our times, and that our times aren't so new.

Maybe it's naive of me to think that white readers will work through defensiveness and continue reading and thinking with this essay. But, I can't write to the worst impulses of white people and think it's going to make a substantial difference in a world that kills daily, in large and small ways, with its racism.

Her approach to race goes further than just the comparison between vaccination and slavery. While Whiteness is underexplored in her work, race seems a consistent metaphor for her. Blackness appears too often in this book as metaphor and not often enough as a living character, as contemporary thought. Very few Black American thinkers are cited among her many sources. She also writes, for example, "While there are many well-documented instances of crowds making bad decisions, lynching comes to mind. . . ."

Does it? Lynching in the Jim Crow South was not an *accident* of *mob mentality*. It was a planned, government-supported spectator sport. When writing about lynching, Biss is writing about Whiteness, our *freedom to harm*. To reduce lynching—state-sponsored racial terrorism—to "crowds making bad decisions" is so ahistorical as to be incorrect. This is an interesting turn, though; Biss is clearly a liberal who thinks lynching is wrong. But again, like slavery (also wrong), American racial terror against Black people is being used as a metaphor to describe current behavior around vaccination.

Lynching is white people as king, as sovereign. Lynching gives white people—even poor white people—that ultimate power: to torture and take life, as long as the people are Black. Lynching is an extreme manifestation of the freedom to harm, not a mistake of mob mentality or the "bad decision" of a crowd. Surely she could make a comparison to bad decisions that doesn't involve race at all, instead of one that invokes White Supremacy this way. For example, capitalism is rife with illogical collective decision making, from the dot com bubble to the subprime mortgage crisis (which disproportionally harmed people of color).

When Biss decides, at the end of the book, to donate her O-negative blood (she's a universal donor), she's surprised to find out that the man also giving O-negative blood next to her has "dark skin." She goes on to consider where O negative blood is mostly found: "it is also," she writes, "somewhat common in people from Western Europe," presumably her, "and parts of Africa," presumably the other donor.

"We are an extended family," she writes of the man with "dark skin." Bile rises from my gut. There's an absence here, a void. Biss never names her own race nor the race of the man with "dark skin." Dark skin is not a race. Her unwillingness to name race even in this moment where it is *needed for clarity* stands starkly against her use of racial imagery (slavery, lynching) as metaphor or example. This is an elusion, a way of looking away from her own Whiteness. If we don't name the race of other people, we don't have to name our own. If we don't have to name it, we don't have to face it in our work.

I'm a queer, white college professor now, non–tenure track, and

I make enough money to live more comfortably than I once did, solidly middle class in New York City. I fuck up on race; I'm a white American, and when I do fuck up, I hope to be called into an understanding of why what I did was wrong and how to be accountable for or undo that harm if possible. That's, I think, what I'm trying to do here, too. I didn't do that work in my FB post; it was a mean-spirited call-out, not a loving call-in. I wasn't writing to Biss; I was venting to my friends. But it was online. Biss did see it (although I did not tag her, some mutual friend must have passed it along), and she reached out to me. She was right to take exception to my post. I wondered, but didn't ask, who'd sent it to her. I apologized for the tone and content of my post. I told her I admired her writing and research, but had concerns about how race was used and presented and elided in the book. And that viewing science through a lens of unexamined Whiteness leads to what my friend Chanda Prescod-Weinstein has termed "white empiricism," which is often incompatible with science's presumed mission of discovery. I don't think I'm wrong to be angry at her book, or to write—lovingly, I hope—about where she—as a white writer—went wrong in my view. And so to learn with and from it, and to hopefully share that learning here.

It was harmful and not useful to post on social media the way I did. I'm sorry for that. But it also feels harmful to use her work in my book and not discuss my critiques of it simply because I don't want to hurt her. Undoing harm—or trying to—is more important to me than any one person's feelings (even mine). And while Biss told me that in the years since *On Immunity* was published, she has engaged with several writers about the book and reexamined some of her decisions in the book, including citing few thinkers of color, it hasn't to my knowledge been addressed publicly. And it should be. That would be a step toward undoing the harm that I think the book did—and does. Patrick and I used her thinking in our essay about war metaphors, and so, here, too, is my problem with her useful book.

....

In 2020, Amy Cooper, a white woman, called the police on a gay Black man, Christian Cooper, who was birding in Central Park. She probably didn't view herself as a racist person, and in fact she said so after the fact, but there she was, endangering a Black man's life in eerily predictable ways. He was *threatening* her, she said.

As T. J. Tallie wrote in the wake of the Amy Cooper saga, we—white people—all denounced Cooper, which puts us all on the right side of race relations in 2020. "How *could* she?" we other whites said.

But Cooper could have been any of us white people. Tallie likened this way white people kill without thinking themselves "bad" or "racist" people to asymptomatic lethality, the fact that people who don't feel ill can spread COVID-19 and lead to deaths beyond their own. He writes, "Like an asymptomatic carrier, white people move through the world feeling that things are alright. Yet interpersonal interactions bring with them the possibility to be very lethal to black people." Among Tallie's points is that both COVID-19 and police interactions are more likely to kill Black Americans than white ones, and that white people are too rarely cognizant of the harm that they can do in a fearful heartbeat or an uncovered breath.

Not calling the cops on a Black man minding his own business is truly the lowest possible bar to jump over, and yet some of us are barely making it. Years ago, at a panel in New York, I saw the writer Kiese Laymon describe how men get kudos for not doing the worst harm. When the bar is on the floor, he said, it's easy to feel like you're flying.

But am *I* flying? And why do I feel the need to be? It's not for cookies, I don't think. Failing at this mission is literally deadly; racism continues to kill, and I don't want to be a murderer. It's worth working every day at that. It's not about cookies and kudos; it's about not perpetrating further harm in my Whiteness and then working to undo the harm that Whiteness has already, and always, done.

In *On Immunity*, Biss is fundamentally writing about freedom: with vaccines, the conflict is between our *individual freedom* not to take a vaccine and the *freedom to harm others* that this choice confers.

The freedom *to* harm? To not get vaccinated, to not wear a mask is the freedom of the sovereign, to be the king of one's own body, even if it directly harms others. Why can we not be the king of ourselves, you might well ask. We can, where it does not harm others. It often harms others; this reminds us how poorly separated we are from one another. This calls into question whether we even have a self, a separate body. This reminds us that the earth is ours, that our health is public as well as private. To attempt the impossible freedom of the sovereign in order to imagine we have control? That is embedded in American Whiteness. We have to accept how little control we do have over our lives and the fact that illness will come for us all.

The thing I'm most proud of this weekend? The thing, the most free thing I've done?

Two answers. I put my body on the street in a protest, a deeply hopeful and loving thing to do, I think. The world may be mostly broken, but at least we're trying to change it. And I made food for my partner and Andrei, my friend. Loved them. Held them. I was free and freely loving. I made Andrei a Negroni. I poured myself some wine. Before he left to go home, Andrei and Devon and I stood outside on the balcony sipping our drinks and smelling the humid summer air, the night summer air, thick and fragrant and free.

....

Since quarantimes, Devon and I end our nights watching Netflix or Prime or Hulu. On Hulu, we get the same ads again and again and again, all night. A Hulu ad: Ancestry dot com, the ad tells us, will help find your family history, like that granddad who fought in the Second World War or that great great uncle who fought with the Union against slavery. The pictures are sepia toned. "Bring your backstory to life." Maybe your great great gran was a suffragette!!! Maybe your great great great great gran was an abolitionist!!!

Devon and I start a running joke, our voices in the tone of the narrator: "Was your ancestor a slave owner?" he asks.

"Did your great great gran fight against women's right to vote?" I reply.

"Did your German ancestor fight in World War II?" I add.

A minority of white people in American history were abolitionists. Half the people who fought in the Civil War fought for the southern states. The reason a woman's right to vote came so late is that it was resisted, and anyway, it was only won for white women, and the argumentation was about solidifying white voting power.

We so want to be the protagonists in our own stories. We want to fight against a clear villain: sexism, racism, Naziism. That's the American story! That's the Greatest Generation!

This is so typical of the mythmaking of nation-states. In France, during World War II, most people either actively or passively accepted the Nazi occupation. A tiny minority risked death to join the Resistance. But after the war, Charles de Gaulle pushed a unified national identity: *La Republique* grew out of *La Résistance*; during the occupation, the great French people fought back. Yes, some did. The true story is that the vast majority did not. James Baldwin wrote, "Not everything that is faced can be changed. But nothing can be changed until it is faced." We cannot change until we admit to our culpability in the death making of our nation. This is not meaningless self-flagellation—something that I know well enough from what's left of my Irish Catholic ethnic heritage. It's truth telling that is necessary, but not yet sufficient, for a national transformation.

This is real life, and there are no protagonists, especially not in empire, in nation-states, in America, especially not one's white ancestors. We white people *all* contain villains within us. My grandfather was typical of his generation and he was a racist, sexist homophobe who loved me very deeply, whom I stopped looking up to, whom I stopped speaking to, who died slowly of dementia, whom I carry in my DNA.

Health might not feel like a privilege, and it shouldn't be. But we have to look at the fact that our nation deprives so many others of it. If health is a right, then what are we doing to ensure it's not just a right for *us*, for now?

....

I'm writing to you from 2020. It's an election year. Just last night, at a Trump rally in Nevada, thousands packed in, and no masks. Herman Cain already died from just this behavior, and yet this behavior continues, and it will continue, to kill.

Some Americans take glee in the fact that not wearing their masks will kill Black and Brown and Indigenous people. It's a new way to do harm in the name of their own freedom. Like in a war, like in the Civil War, it's worth risking actual, material death in order to be free to harm Black and Brown and Indigenous Americans. The simple tool of health, of disease prevention, a mask, becomes a symbol of oppression.

Our choices aren't about the harm they do to *others*, but the freedom they grant to *us*.

Protecting ourselves and others from COVID-19 is another way of saying—through action—that Black Lives Matter, Latinx Lives Matter, Indigenous Lives Matter. My risk is yours, and yours is mine. We are more a nation now than ever, bound together, but more than a nation, a globe that breathes air, and air respects no borders.

Whiteness won't save us from COVID-19. Whiteness has only succeeded in what it always does. We aren't suffering less, but other people are suffering more because of us. That COVID-19 can and does kill white people who go to Trump rallies doesn't matter as much as the fact that it will always kill Black and Brown people *more*. This ideology, this Whiteness, is deadly to white people, too. What won't we do to maintain our right to harm others if we are willing, even today, to die for it?

Early on in the COVID-19 crisis, we learned that SARS-CoV-2 damages tissues besides the lungs. In severely ill patients, the virus can lead to multi-organ failure, as the circulatory, nervous, and digestive systems all shut down together.

But subsequent research showed that even mild COVID-19 cases cause measurable damage to the heart. The heart expresses the virus's receptor—the Ace2 protein—and so it's likely that the virus directly infects and harms the heart. In one small study, 78 out of 100 COVID-19 patients had myocarditis, an inflamed heart, after their infection, and the severity of their respiratory symptoms made little difference to the impact on their heart.

White people—young and old, high- and low-risk—are breaking our own hearts when we refuse to wear masks, when we gather, when we eschew the risks because we might be younger or otherwise healthy or because we have the best healthcare on earth, or so we imagine. Racism harms white people. Here is material evidence.

Whiteness, as I imagine it, is an attempted buffer against the worst pain of living. Whiteness is an attempt to experience the opposite of loss; we don't bury our 12-year-olds after they are shot by cops; our mothers aren't dying in childbirth; when we have pain, we are given pills to ease it.

I want to extend this, though, to the moral aesthetics of WASP culture, which I think Whiteness and white people aspire to. Think of the most traumatic and most human event of all: death, the one thing all lives will share.

In WASP culture, it's inappropriate to wail out loud, even at the death of a beloved. This is what Muñoz meant when he called the aesthetic of Whiteness "minimalist to the point of emotional impoverishment." We cry at the funeral, are stoic at the wake, and are back to work by Monday. Capitalism hums along, paying us for our labor without disruption. The neighbors might bring us food, an acknowledgment that it may be too painful to do the simple act of feeding ourselves, but they don't stay to eat with us.

Feelings can fill us up, but only so much. Women are hysterical if they feel too much, and so crazy, and men are feminine if they do, and so weak.

Wouldn't it feel good to shout out to the heavens with joy or with pain, to sob because the one thing all lives have in common is loss, suffering, pain, death?

If we don't see other people's grief, we don't know what it is to grieve. The first time a friend died. The first time a man I lived with packed up and left me. I was unprepared for the way that grief took over my body entirely. Sleep? Work? Ha. What was all this? Why had no one told me? Would I have wanted to get older if I knew this was waiting for me, this pain, this grief, a fact of every life lived?

In this way, patriarchal Whiteness harmed me, and always has. I've always felt too much, messily and in public.

If Whiteness is meant to be a buffer against suffering and pain, anything painful must be a conspiracy against us personally and white people generally. Whiteness won't protect us from COVID-19, but we grow up white without imagining our bodies can be as vulnerable as we now find them to be. The response is denial, and of course it is. How painful it is to suddenly realize that death and suffering belong to you, too. This inability to feel pain and accept it and learn from it makes us unable to live well, even prior to the pandemic. In the pandemic, it has made us murderers by small acts, killers by thousands and thousands of mask-free indoor dinners.

James Baldwin wrote, "I imagine one of the reasons people cling to their hates so stubbornly is because they sense, once hate is gone, they will be forced to deal with pain." We are risking embodied pain, the breathlessness of COVID-19, in order to imagine ourselves as invulnerable as Whiteness taught us to feel.

Whiteness, then, is a death cult, too. Pain, death, suffering, and loss, these are what it means to be human. No, not always. But sometimes, for everyone. If we're not capable of living with that certainty, we're not capable of living at all. I sing my suffering from my windows, I post it online, in order to deal with it, and live beyond it, in order not to inflict suffering on others because I'm feeling it myself.

If Blackness in America always implies a closeness to death—at the hands of a virus or a cop—in order to make white people feel further from death, then Whiteness in America has always implied and still implies acceptance of this death, a quiet way to murder. The freedom to harm is often invisible: it's not always just white people lynching, punching, spitting on. It can also be indirect: the absence of a mask, voting against voting rights or universal health insurance.

I don't want to murder, I don't care how quiet it's kept, how easy it can be to look away. I don't want to look away.

What can white people do in the struggle against White Supremacy? Well, we can read all the essential books that undo the partial history of the world we were taught. Cedric Robinson called the history that white nations teach a "structured ignorance." White people do not know our own true history. We are taught not to know the connection between violence and White Supremacy and our history and identity because if we saw it clearly, at least some of us would be repulsed and want to change. We can read this anti-history and anti-art that somehow managed to get made against great odds: Baldwin and Morrison and Robinson and Fanon and Foucault and Ignatiev and Roediger and Muñoz. It seems that with each new crisis of Whiteness, with each new Amy Cooper, white folks buy a bunch of books they don't read. For whatever reason, I don't see Fanon or Robinson on those lists very often, perhaps because they'd make their white readers and reading groups uncomfortable. They made me uncomfortable when I read them, but I also see their inherent truth in my own life: white masculinity harmed me too. Reading Cedric Robinson and Harold Cruise changed my relationship not to Blackness, necessarily, but to *my own* race and history.

White people can and must listen to Morrison and Muñoz when they tell us that white people have a very serious problem, that white people have violence baked into our cultural identity, and that it's white people's job—alone—to fix it. It isn't Black or Brown people who are emotionally excessive; it is white people—trust me—who are emotionally deficient. White aesthetics and emotional minimalism do not prepare us for the pain of a life lived. "Leave me out of it," says Toni Morrison. Capitalism, patriarchy, homophobia, White Supremacy: these things harm us all, they limit all our potential, they make us human only if we're *better than*, which is no way to be human at all. Many Afro-pessimists don't believe white people are capable of changing, undoing our systems of power. That they may well be right doesn't mean we should slip into cruel nihilism. We must try. Not to save people of color or Black people, but to save ourselves.

"Everyone talks about dismantling structural racism," infectious disease doctor Bisola O. Ojikutu said in a webinar on COVID-19 and health disparities. "Dismantling structural racism will require moving around power. And people don't want to do that." We need to be willing to do that. We need to lead the way on giving up power. We need to make a world where power isn't our driving force and influence isn't zero sum.

We can, and must, organize. Our workplaces. Our buildings. Our blocks. Focusing on our white partners, inviting others to participate if they are able. We can make pledges not to call the police and to invest in decarceral mechanisms of conflict avoidance, like transformative justice circles. We can build mutual care networks, sharing resources with those in our community who need them. We can organize to help one another vote and elect representatives who reflect our anti-racist anti-capitalist values. We can and must work to divest from capitalism itself, or we will never, ourselves, be free.

I don't want a COVID-19 broken heart. I don't want to harm others to feel at less risk of harm myself. I don't want to pretend life won't hurt; I know that it will. White people, and white men especially, need to grow our emotional muscles so that when we feel pain—a guarantee in life—we can feel it, accept it, learn from it, hurt, and heal, instead of attempting to cut short our suffering by inflicting pain on another. No longer.

White people have a very serious problem. We are murdering others and the planet in order to feel powerful, to have power over. We force each other to limit our emotions to the point where we cannot live a full life without anger and hatred and vitriol. The only way out is through. We must turn our anger inward, and toward our past, in order to immolate Whiteness itself, to destroy it entirely. There is no possible reform for an identity with violence at its inception and core. Whiteness harms white people, it just harms other people more. This is not freedom. I don't mean we must plan to kill white people. White people have a very serious problem. I mean killing Whiteness so that white people can finally live free.

9

On Activism and the Archives

The Looking Is Always Worth It

2013.2020.mem.001: February 13, 2013 (A Memory)

It was Valentine's; at least we'd fuck.

Whether I had a cheri(e) or not, I always hated the day, for its red roses, for its consumerism, for the fact that everyone seemed to notice that I had a passing resemblance to cupid, all rosy cheeks. That year I had a sort-of boyfriend, I called him *my cheri.* Kaliq wasn't much for Valentine's either; he was more about public sex than public displays of romantic connection, and he hated pink. He could be soft and kind and romantic, yes; only in the house. He loved my cooking, and so I'd cook. Something special: steak, spinach salad, even though I preferred arugula, and my roast butternut squash with maple syrup and chipotle, his favorite.

I don't have a journal from 2013, but I know the night exactly. Valentine's again? Please no.

What to do with a night we'd be locked inside, hiding? We'd been meaning to watch the movie for two months already, and I'd had a DVD—remember those?—for the last month or so. I was teaching at Vassar that semester, biochemistry. In this course, students learn fairly complicated math around inhibitors of enzyme function, essentially how almost every drug humans have ever invented or found works. Based on this math alone, you can more or less figure out how the drug is working: competitive or allosteric, weak binders and Trojan horse inhibitors.

Cocaine is a small molecule inhibitor, and so is the Lexapro and finasteride I take nightly. Advil and Tylenol, too. Most of my students wanted to be doctors. If you don't understand Michaelis-Menten, I yelled at them, you shouldn't be allowed a prescription pad.

To drive home how the material we were studying was applicable to saved lives and public health, I wanted to show a film about drug discovery. The year before, the film *How To Survive a Plague* came out with archival footage of ACT UP New York, focusing on its role in developing the drugs that changed HIV treatment in 1996, saving so many lives. I'd asked the Vassar library to order a copy of the DVD so I could watch it, and if it felt right, share it with my students. Kaliq borrowed his roommate's PlayStation and connected it to the TV on his low bedroom dresser. We brought plates of food into his room, lay on the floor, our food in front of us. I leaned up to kiss him.

"Can I start?"

"Of course, eat while it's hot."

I kissed him on the cheek.

"I meant the movie," he said.

I bit into a forkful of steak I couldn't afford.

What to say? What happens next? Larry Kramer shouted about a plague, a bunch of fags didn't want to take it anymore. A bunch of queens fought as hard as anyone had ever fought. They fought to live. The film tackles the science advocacy of the ACT UP Treatment and Data working group and the Treatment Action Group (TAG) that grew out of it in 1991; the main characters, Peter Staley, Mark Harrington, Bob Rafsky, David Barr, worked alongside and learned from chemist Iris Long to become world experts in the virus that was, even then, killing so many in their community. They refused to die quietly, protesting and shouting, insisting that "Silence = Death."

ACT UP was multiracial and multifocal, from homelessness to gentrification to sexism in HIV treatment, in addition to drug access and science. *How To Survive a Plague* is only one ACT UP story, one with mostly cute cis white men who—prior to the plague years—had mostly studied at good colleges or worked on Wall Street. Even upon first viewing, this much was apparent. And yet, the story the film told—as a queer scientist, as a gay man who grew up in the early plague years—hit, and it hit hard; it knocked the wind from my chest.

I chewed my food—did I taste it at all after all the work I put into it?—but I wasn't there, with him. I was then. After we ate, he stood up and turned off the light, and there was nothing, no food, no room, no him, almost, there was nothing but the then. I held his hand, Kaliq, and he held mine, only his hand to tether me to the world, the now. From the Ashes Action, when those who lost loved ones to AIDS brought their ashes to the White House and dumped them over the fence, laying claim that they didn't die, they were murdered, that their lives mattered then and their deaths mattered now, to the end of the film, I cried. One man threw ashes over the fence, shouting twice, "I love you Mike!" He and two other men hugged and cried then, holding one another, even as the cops moved in, on horses, wearing gloves. I knew who I wanted to be like, who I wanted to be. I wanted to love and cry and hold and fight.

ACT UP and TAG pushed scientists to work harder, faster; they pushed for exhaustive testing of the best drugs as fast as possible. They insisted that drug companies invest in finding treatments, too. Ultimately, researchers did. They developed protease inhibitors that block a required viral enzyme and so block viral replication. In combination with older drugs that blocked other viral enzymes, the protease inhibitors brought people back from death's door. People, in 1996, started living.

I sobbed. For the living, for the dead.

The next week, before assigning the film to my students, I told them that my boyfriend and I had watched it on Valentine's Day, not exactly the most romantic. They laughed. I'd never come out to them officially, not that I had to. But I wanted to; the film made me a little bit braver.

This story could have so easily been lost. Credit is due to the scientists, but not only to them. They were pushed, prodded, helped along the way by activists. How do we know? Because, at ACT UP meetings and demonstrations, protesters brought cameras. Because ACT UP included artists who made signs and posters and flyers and activists who saved them so even I—decades later—on a museum wall, on the Internet, could see (SILENCE = DEATH; AIDS GATE, bright yellow, Reagan's eyes bleeding pink; THE GOVERNMENT HAS BLOOD ON ITS HANDS, a bloody hand print; CARDINAL O'CONNER: PUBLIC HEALTH MENACE; AIDS IT'S BIG BUSINESS—BUT WHO'S MAKING A KILLING?).

I knew some of this, but not nearly enough. I knew that in the 1980s, activists fought against government inaction toward the AIDS crisis. I knew that many of these activists were queer and I knew about ACT UP. I didn't know that the meetings were so loud, that people disagreed so much, that fags stood up to harassment and arrest, that they became experts. I didn't know *why* Bill Clinton said "I feel your pain" in his Arkansas drawl. I didn't know until I saw.

"Shit," Kaliq said.

I said nothing. I stood up, red eyed, and turned on the light, and brought our dishes to the kitchen, turning the sink on, the water too hot. I finished the dishes, my hands now red, too, and stinging.

"I didn't know so many of those things," I told him later.

"Me neither. I didn't know those people either. Peter whatever and all that. Can you believe?"

I could, but I couldn't.

Later that night, I reached for his cock in bed, but he said he couldn't—babe—not after seeing that movie.

"But it's Valentine's!" My voice was soft but surprised, a little break in the phrase.

"After that? You still wanna fuck?"

"Of course. Feel." I was hard.

"Sorry I just can't," he said, "I'm too . . . I don't know. It was a lot."

It was, and I almost understood. I turned over and threw my arm around him. Accepting it, he breathed heavy and slept, my own sleep not far behind.

....

2021.2021.expo.001

Among my favorite work by Félix González-Torres are his puzzle pictures: photographs that he prints and then cuts into individual puzzle pieces to be put together by a viewer. Most of the pictures you can find of these photographs online show them assembled, a faint crack in the surface making their incisions apparent. I prefer them in the rare images one can find, pieces collected in a clear plastic bag. The photo could be anything. I long to be the one to piece it together.

A puzzle is like an archive to me: pieces of information in no specific order that must be placed against one another to see what fits. A story that comes together only by searching and testing and looking some more. A meaning that is made by the person who left the information behind and the person building the puzzle in equal measure.

González-Torres died of AIDS. His archives are kept by the Félix González-Torres Foundation and includes myriad images of his installations and works, oral histories of the installations by those who helped build them, his work archive, over 100 interviews with people who knew Félix. The interviews are still being collected. The Foundation also keeps an archive of his correspondence, which acknowledges that "the Foundation honors that these materials belong to the individuals who are the recipients, and respects that each individual recipient has the right to determine if the material is best kept as part of their own archives or otherwise disseminated."

González-Torres's puzzle pieces feel like his body, in a way, still living; touch something he touched, build something he made. Live on, together, with this object, together. The meaning of this image is nothing without him, and without me; only then does it become clear.

2020.2020.journ.001, Thursday, April 16, 2020 (Journal Entry)

Isn't it weird that sometimes I can't open this doc, and other times, I run to it at the end of the day, to write, and so to be sure I remember.

Today, working on NYU exam stuff and on a document about serology tests for COVID-19 for the Working Group, why we need them, and how to make sure we get them. No. My God. I forgot about a Zoom with Mark Harrington and Wafaa El-Sadr, and they email me to remind me, and so I hop in. When I got invited to work with this group, to do COVID activism, did I know I'd be working with these people? No, I just showed up. We speak for only 30 minutes. All our videos stay off. I know what they look like: Wafaa in her office, hair up, looking tired but sharp, Mark with a virtual background—New York's skyline. I know their voices so well by now. I tell Wafaa that I'm already drafting a document about the importance of serology tests, and outlining an op-ed (for someone else to write, someone more famous, someone with scientific authority and name recognition) is not much extra work, don't worry, I can do it by the end of the day.

She says, "Great, that's great," in a tone of voice that I can tell she means it. It sounded like relief that something will get done and that for once she wouldn't have to be the person to do it.

We're all tired.

But my brain seems to be working just fine, thank God, for now.

I finalize the exam stuff so that after our COVID-19 Working Group Diagnostics and Treatment (CWG D+T) call at 5:30, I can start cooking. Fish tacos tonight, with guac, pineapple salsa, lime crema, tomatillo salsa. Cooking helps. I can taste the things I make, I can run my knife through cilantro and for a moment my whole world is a sharp blade and the smell of that herb.

The call isn't at 5:30 though, but 5:45. I'd forgotten. I'm using my Zoom account to host, so I hop in early and have 15 minutes to waste before others arrive. I turn off my camera, mute my mic, and Teebs texts saying he's in performance artist Justin Vivian Bond's happy hour on Insta, and so I go there too, and I love Justin, today in a full platinum blond wig, holding a drink, holding court. Devon comes in and watches with me, pressing his body against mine, and I close my eyes.

"Yes darling," Justin says, "we didn't call our drink an Aviation today because none of us are flying, dearests, we're all here on the ground. And the government tried to give all that money to Boeing, didn't they, honeys, but Boeing said no, didn't they, because of the STRRRRINGS attached. Boeing didn't want the strings, babies, they wanted to take our money but not have to tell us what they did with it, and Nancy Pelosi didn't like that, no she didn't."

Devon is smiling, and I'm smiling too.

On the 5:45 call—now—there are three people who'd built ACT UP and TAG in the 1980s and 1990s—Mark Harrington, Garance Franke-Ruta, and David Barr.

Mark, Wafaa, and I talk about the serology work. We're trying to get all the labs in NYC and the metro area who are developing and testing antibody tests on their own to speak to one another, to share reagents, to collaborate and use the best test instead of their test, if they can.

Garance looks sideways at the camera.

"There are just so many questions we don't know." Silence, then. "We don't even know how long people who get sick can transmit the virus. That's like . . . quarantine 101. How long to quarantine for. They're just telling people like seven days, and oh, go back to work," says Garance.

"This is giving me 1980s flashbacks," she adds.

"I'm incensed," says Mark, and you can hear it in his voice.

"Same," says David. He was there too, in the 1980s and 1990s.

"I'm watching," I say to myself, "PTSD in real time."

I look at my own head shaking in the Zoom window. The lighting in this room is bad, straight and direct down from the ceiling at Devon's place.

"We need central coordination," I say, "of research activities. A spreadsheet of the questions that need answering and who's working on them."

We can only do exactly as much as we can do. Today's action items: Write a list of the biggest questions we need answers to, now. Find out if they're being studied. Insist they be. James has been working on a document for a week now, mostly about PrEP for COVID-19, whether a pill might be able to prevent infection. But there is no drug development pipeline. There is no coordinated drug screening looking for drugs that stop COVID-19.

"I just realized what a big problem this is," Mark says.

"And what a big problem it will soon be for us all," I say to myself.

....

2017.2020.mem.001, Friday, February 10, 2017 (A Memory)

A chilly night in Washington, DC, yes, but warmer than New York, at least, I told myself as I wrapped my neck in a long orange scarf. I'd traveled down, on the train, for a writer's conference I attend every year. That's how I know this date too. I splurged on the expensive train over the Chinatown bus because, in DC, I had a free bed to sleep in. Every time I traveled to DC, I stayed with Jesse.

Just before I visited, Jesse purchased an apartment, the first of my friends to own anything. He'd continued working as a scientist, but did his postdoc in industry, working for a small pharmaceutical company in their research program, which starts at double the salary of a research postdoc at a university. Since I met him, he'd grown a thick beard, one that—and I was jealous of this—managed to connect all the way to his hairline. DC was cheaper to live in than NYC, although prices were rising. But he'd lived mostly outside DC, closer to work in Bethesda, saving money. Now, he owned a two bedroom in Trinidad, a residential neighborhood, still mostly Black, in DC's Northeast.

From Wednesday until Friday, I'd come and gone from his house and felt bad not spending time with him. I did see him more often than the writers I knew at the conference, but I missed him more, and we needed at least one night out together. Friday night, then, we did dinner, just us. Then we met some of his friends for drinks.

We walked up the stairs of Number 9, the bar Jesse drove us to, Jesse first, and I could hear music growing louder and feel the temperature of the street, cold, fall away and the temperature of bodies moving replacing it. Following Jesse up the stairs, I saw the room inside, long and narrow, a bar to the left, a hallway to the right, and a little open space right at the top of the stairs. His friends were waiting, smiling wide when he walked in, hugging me for the too simple fact that I was with him.

It was 9 p.m. I was not-quite-drunk and in a room full but not overfull of homosexuals. DC's gay scene is even more segregated than New York's; in New York, a lot of bars are mostly white, a few are mostly Black, and a few are relatively mixed, at least on some nights. In DC, most of the gay bars that had more than a Black person or two had mostly Black people, and the white bars were hostile. Jesse's friends, the ones he went out with in DC, were mostly Black and mostly gay, and so—when I joined him out—we went to Number 9 and most often Cobalt, which closed in 2019.

I grabbed drinks for Jesse and me, gin and tonics with Hendricks because we weren't as poor as we used to be, and brought them back to him at the open area near the front of the bar. The light inside was purple, brighter than a nightclub, easy enough to see who you're talking to.

"Hold this, I gotta go piss," Jesse told me.

"Sure, no worries," and I put it down on the table in front of me next to my own drink.

"Wait, how do you know Jesse?" his friend asked me.

"Mutual friends, we met ages ago. But we got close cause we're both scientists."

"Oh shit, you're a scientist too? Another nerd who hides it in skinny jeans," he said, laughing.

"I try, but I give myself away as soon as I open my mouth."

"Just like Jesse," he said.

"Just like him," I said.

I said hey to everyone around the table, not exchanging names, just a word and a nod.

Conversation shifted back to mostly one on one, and I drifted to the wall, where I always feel more comfortable, even in groups of people I know.

I stood next to one of Jesse's friends in a burnt orange coat, and I asked him where he got it. Easy conversation moved from there; I don't remember what, gay things probably, like shopping and Britney and Whitney. A friend of this friend laughed with us too as we talked. I'd introduced myself to Eric—orange coat—already, and so I turned to his friend. He had short cropped hair, a nose ring, and red-framed glasses. He was shorter than me, handsome, with a big smile and eyes that said more than his mouth, even already.

"I'm Joe," I said.

"Steven," he said, smiling.

"And what do you do, Steven?"

"I'm an archivist."

....

2020.2020.journ.002, March 2, 2020 (Journal Entry)

Today was our first call. James Krellenstein, who cofounded PrEP4-All, DMed on Twitter to ask if I wanted to join a science advocacy group of HIV and other activists focused on COVID-19. I didn't know James, but I knew his work with PrEP4All. I knew we had to do something about COVID-19 because it seemed like no one else was.

Why was no one else doing anything? I'm just a molecular microbiologist, I don't study coronaviruses. I don't have a public health degree. I knew just enough to know that this was about to be very bad, that in fact it probably already was.

The call was introductions and ideas. Will we work mostly on federal stuff or stuff specific to New York City and State? It feels like New York is where we'll have the most impact. What should we be pushing for? Nonpharmaceutical interventions: social distancing and an immediate shutdown of just about everything—schools for starters—until we can get testing up.

And why can't we get testing up?

We set up working group calls, including Diagnostics and Treatment, D+T, on Tuesday and Thursday at 5:30 p.m. In the early HIV crisis, there was a committee that focused on these issues as a part of ACT UP. That committee eventually split and became the Treatment Action Group, TAG. TAG is a founding member of this new group. On my Zoom screen there is Peter Staley, Mark Harrington, David Barr, Garance Franke-Ruta. There were people I didn't know of yet, but would soon: Wafaa El-Sadr (doctor and professor of epidemiology at Columbia) and Charles King (head of Housing Works, an organization that works with homeless and HIV-positive populations), C. Virginia Fields of the National Black Leadership Commission on AIDS, and Guillermo Chacón from the Latino Commission on AIDS.

What do we call ourselves?

Someone, I don't remember who but I think it was Peter Staley, suggested the COVID-19 Working Group, and we all agreed, nodding heads from all over New York, all across Zoom.

2020.2020.journ.003, Friday, March 13, 2020 (Journal Entry)

I'm on a call with Mark Harrington and James, just us three. We talk for about an hour without a specific agenda. We brainstorm about testing. Mark and James have contacts in different government agencies.

"Labs still can't do anything," James says, "they can't set up their own tests. We have no idea what's going on. But the CDC and FDA won't let them."

We talk then about what we should be doing, what tests we should be making, and how they should be used.

And how can we know how widespread the virus is if we can't test? So, I talked about the research in Seattle that used RNA sequencing to determine the size of the outbreak there. If you sequenced two viruses, you could determine if they're related. If two viruses are related, they're most likely descendants of one another. If you have a descendant of a viral infection in a population, that's clear evidence that the virus has been spreading in the community for at least weeks, probably longer.

If we didn't get the city to shut down soon, lives would be lost. We knew those lives were more than likely to be mostly Black and Brown, mostly essential workers, and maybe healthcare workers, many of whom, in our city, were Black and Brown too.

"We could do that here," I said.

"I think we might be able to get the samples," James said. "Let me reach out to some folks."

"Talk to you Monday," we say to one another at the end of the call. "Have a good weekend."

I texted Ngofeen, "We're gonna try to get the RNA samples and sequence them."

"What does that mean?"

"It's too much to text . . ."

He'd just started a new job reporting for *The New Yorker* podcast.

"Can I call then? Honestly, let me just record it. It could be a good story. And even if it's not, if we don't record tape now, then we'll never have it. If we record it, worst thing is we don't use it."

2020.2020.trans.001, Sunday, March 15, 2020 (Transcript of In-Person Interview with Ngofeen)

Ngofeen Mputubwele: Can you tell me who you are?

Joseph Shannon Osmundson: My name is Joe Osmundson and I'm a biologist. I teach biology at NYU. I did my PhD on viruses and bacteria at the Rockefeller University here in New York and in my postdoc I studied yeast.

NM: You sound really tired.

JSO: Fucking exhausted. And so, yeah, I mean, what's been happening is that scientists have been watching this thing. When I took virology in grad school, the one thing every lecturer said was that in our lifetimes, there will be a global pandemic that affects public health on a global scale leading to hundreds of thousands to potentially millions of deaths. What happened in Wuhan is terrifying. The hospitals were quickly overwhelmed and the virus seemed to be out of control. And so we were very worried at that point about this virus and its potential to wreak havoc on healthcare systems.

Hang on. I'm being a bad host. Do you want a water? Seltzer?

NM: Do you have a seltzer?

JSO: I'm such a fag; I need seltzer water for the apocalypse. Honestly, this is the last seltzer water I have in my house, and if the store downstairs is ever out of seltzer water, I'm going to blame you for taking my last one before I kill myself because there's no more seltzer water. Let's try not to joke about suicide on the radio.

[Ngofeen pauses to check audio equipment.]

NM: So we're going to talk about a few things. But first off, just like where are we physically and where are we emotionally?

JSO: Physically, we're in my apartment. And I have been in my apartment basically for the last two weeks straight. I live in Chinatown in New York City.

One of the scary things about this virus is that makes it spread so quickly because so much of the disease is mild. Especially if you have no testing. And we have no testing. That's the problem. It's March what? 15? We think this is serious enough that we need to close schools, close bars, close restaurants, stop having large group events. We literally have no idea how many people in this city or in this state are infected with this virus. We are completely, completely unaware.

But there are ways we can detect it, or at least estimate.

NM: So you were telling me that some kind of results just came in. I just came into your apartment at Sunday, March 15th, and you said that some kind of results about New York City.

JSO: Let me tell you the story that happened in Seattle, and that is what we're trying to replicate in New York. So in Seattle, actually, this virus hit very hard in terms of death because the first places that it was really infecting were nursing homes, which have very critical patients for this virus. So dozens of people died very rapidly.

NM: And you're from Washington State, right?

JSO: Right. So this virus literally shows up in the United States in the county that I grew up in, which is the middle of nowhere in Washington State. And currently, the biggest outbreak centers are New York City, where I live, and in Washington State, where I'm from.

And at this point, basically, the number of tests in the U.S. was very near zero. There was a project called the Seattle Flu Project that was ongoing there. And they would go out into the community and just test people who had symptoms of flu for influenza just to monitor all the time the level of flu in the community. What the researchers realized is that they might be picking up in their samples people who were negative for flu but had the coronavirus.

But technically they were not allowed to test for the coronavirus. This is a study on flu. So initially they couldn't test for the coronavirus due to ethical issues of informed consent, but eventually they end up doing so against the rules. And in the first cohort that they test, there was a positive patient for coronavirus. This is . . . bad. Imagine it's a population experiment. You have a thousand people in the population who have cold or flu-like symptoms. You pick up five at random. This is just some person in the middle of nowhere who happens to have COVID-19.

Scary as fuck.

What was brilliant was that the researchers didn't stop there. What they did is they took the RNA, and they had maybe three to five samples at that point, and they sequenced the entire genomes. That means sequencing every base of RNA that the virus has. And this allows you to sort of count differences between individual viruses and say like, how long ago did these two viruses diverge? Are they related? Did they come across in the same wave or did they come across in separate waves? And when they did that, that completely changed the name of the game in Seattle. Two viruses in those samples were related to one another. They almost certainly had been spread in Seattle, not abroad. So the virus had been there, undetected, for weeks.

NM: And so that was the situation in Seattle. Make the bridge to New York.

JSO: Yes. So New York, we're in the same case. We're still not really testing. We think we have lots of community transmission that we're not detecting because we're still not testing widespread enough.

Basically, the sequencing for this virus can tell us three critical things. Number one, the approximate size of the cluster that these cases come from, roughly how many people in the community are likely to be carrying this virus. Number two, roughly how long the virus has been spreading in the community.

And three, we can tell roughly how many initiation events arrived in NYC. So we have spatial resolution around the globe, temporal resolution in New York City, and an approximation for the number of people here who are likely infected. It would change everything.

NM: So what happened today?

JSO: I've been working all weekend with Elodie Ghedin to get set up so we can sequence the samples. And she's done it, we got a collaborator to sequence them, and it's not perfect but it's good enough. We'll do it in her lab too. And we can do it now, fast. So on that side, getting the experiment done, I know we can. And we have three labs ready and waiting to go, just to make sure one of them succeeds.

Basically, the COVID-19 Working Group reached out to the city. And what we wanted them to do was just to preserve the COVID-19-positive RNA. They have dozens of samples dating back to early February. Those old samples are super valuable for this type of experiment, because they're more likely to be a "parent" of viruses we pick up later. The proposal that I'm writing will allow us to get the samples in hand, hopefully in less than 24 hours.

Once we have the samples, we can sequence them in a day or two as well. So in a couple of days, we could know approximately how many people in New York likely have COVID-19.

Oh shit, what time is it?

NM: Almost 5:30.

JSO: Shit. What to do? I'm in my fucking apartment. I've been in my apartment for a good jillion hours without leaving and I'm about to hop on a phone call to James. We have been meeting on the phone daily. It's really tough to do activism when you can't put bodies together. It's like a lot of these folks are ACT UP veterans.

And, you know, the idea with ACT UP is like, let's all gather in a room and share ideas.

And let's go do die-ins in a church. Let's go to the mayor's office and knock on the door and be like, let's do a sit-in. "Shut down the city or we're not leaving," and the press will cover it and at least your message gets out there. We can't do any of those things because any event that puts bodies in the same room is exactly what we're trying to advocate *against*. We are trying to model the behavior of being separate from one another. Thank God this is happening in an era where we can do Zoom meetings and we can do conference calls and stuff. So we're about to hop on a call. Time is up.

2020.2020.trans.002, Monday, March 16, 2020 (Phone Call with Ngofeen)

Joseph Shannon Osmundson: OK. Well, I'm out of breath because I was out on a run. And we've been waiting all day. We submitted our grant documents this morning. And it's been radio silence. So we've been kind of on edge all day. And I was just out for a run to kind of work out some of that energy. And I got a message from James to check my email. So that's what I'm going to do now.

Why won't my damn email load?

OK here we go. This is from the lab director. "Hello, everybody. This project sounds great. We will just need to get some paperwork in place before we can start. I'll get this started on my end. How much volume of extracted material would you need?"

They are just doing the paperwork to give us samples. Oh, my God. But you know, you never know until you have the samples in hand.

Ngofeen Mputubwele: How are you feeling?

JSO: It's just really overwhelming. I'm happy we can do something, anything at all. I was feeling so hopeless.

It just feels like all you can do is sit in your apartment and wait for things to get bad. And then for things hopefully to get better. And all the scientists I know, we all feel helpless. We just look at the data and look at the data. Then just three days ago, a few of us on a phone call had an idea. That was Friday and it's Monday now. We worked it out. We worked really hard. All weekend. We worked with a lot of people. You know, it just seems like it actually matters. I didn't know that it was going to.

Now we need to get the samples in hand. We need the samples in our hands. Bureaucracies can stop that at any point. So, you know, hopefully it'll really help a lot of people. Hopefully it really can really, really help a lot of people.

2020.2020.journ.004, Wednesday, March 18, 2020 (Journal Entry)

No samples. No word. No shutdown.

2020.2020.journ.005, Friday, March 20, 2020 (Journal Entry)

Nothing.

2020.2020.journ.006, Sunday, March 22, 2020 (Journal Entry)

NY State on pause. NYU remote. I go into my office to get what I'll need. I don't know when I'll be back. My plants will probably die.

No samples. We've been pressuring de Blasio for weeks. Of all the things that helped us get here, to a shutdown, it was everything but RNA sequencing. We didn't help get here.

2020.2020.journ.007, Monday, March 23, 2020 (Journal Entry)

There are three sequenced viruses from New York. We sequenced none of them; researchers at Mount Sinai did. None of the viruses arrived directly from Asia. They're from Europe. None of them are related. We can't say anything about them yet, we need more. We could sequence two dozen in two days. We don't have our samples yet.

2020.2020.journ.008, Wednesday, March 25, 2020 (Journal Entry)

We still don't have the samples.

2020.2020.journ.009, Wednesday, April 1, 2020 (Journal Entry)

We still don't have the samples.

....

2020.2020.trans.003, Monday, November 9, 2020 (Zoom Call with Steven D. Booth)

JSO: Let me start the recording . . . Hi love! You look good! Oh, what does she have on?

SDB: This is part of my work wardrobe, it's nothing but overalls.

JSO: Let the archival record show that I'm wearing a tank top.

SDB: Joe's uniform.

JSO: Yeah, how are you, my love? I'm just pouring some wine, are you drinking?

SDB: I am drinking, but I'll probably need to go get more after this. For some reason I never buy enough wine at the grocery store.

JSO: I don't have that problem.

SDB: Oh, you don't now?

JSO: I went to the Trader Joe's wine shop today. I got two boxes of rosé, and each box has four bottles in it, so. . . .

SDB: Oh cute!

JSO: Very economical. And it's actually a good rosé. I mean, I buy the most expensive boxed rosé.

SDB: That box that you bought that time when you were here, probably?

JSO: Yes!

SDB: Yeah. Yeah. OK, that's what I'm going to do. I mean that's what it is for the winter.

JSO: I want to start by just asking you about the night we met. What's your memory of that night?

SDB: So it must have been a Friday evening? Yep. And the space we were at with No. 9, N-O period nine. It's not even really like Dupont Circle, but like near Dupont Circle off of 14th Street. I think that's 14th and P actually. And so I was there with two of my friends. One of them had on a bright orange coat. And for some reason I don't believe that it was happy hour. I think we maybe met up for dinner, then came out because the bar was not crowded, which means that the two-for-one special probably had ended. We were upstairs on the second floor. And you walked in with two individuals, and one person from each group and they knew one another. And so, I guess it was just random conversation.

JSO: What were you drinking?

SDB: I was probably drinking Deep Eddy's grapefruit and tonic. That must have been my drink then. I don't even remember there being introductions. Somehow we just started all having random conversations, and I went to go get a drink, I sat down at the table, you were sitting down. You had on a black T-shirt and a baseball cap that was backwards.

Somehow you must have asked me, "What do you do?" And I said, "I'm an archivist."

And so we struck up a conversation, and at some point you all left because I remember there being goodbyes. We stayed a little bit longer and then we ended up seeing you all at Cobalt, which was on 16th and R. And then we danced the night away.

JSO: So before I do a big pivot back to the archives, I'm feeling a little emotional and nostalgic about all this. I mean, one of the things I love about faggots is that you can be out at a random bar and oh, a friend of a friend of a friend of a friend that you meet creates like a bond that last a night or also year and maybe even a lifetime. And you learn from each other and it's like beautiful and I want to cry and shit.

SDB: I think you're absolutely right. That's how I met my best friend Eric, the one I was there with, so I totally understand it. Eric was wearing the orange coat and Christopher was the tall, light-skinned guy.

JSO: That's so funny. And it was Brian, it was Jesse's friend Brian, I forgot that we were there with Brian. And Brian knew someone in your group?

BDB: Brian knew Christopher. Yeah. Yeah.

....

2017.2020.mem.002, Friday, February 10, 2017 (A Memory)

"I'm Joe," I said.

"Steven," he said, smiling.

"And what do you do, Steven?"

"I'm an archivist." He had short cropped hair, a nose ring, and red-framed glasses. He was shorter than me, handsome, with a big smile and eyes that said more than his mouth, even already.

"I just got more than a little hard."

"What do you mean?"

"I love archivists!" I said a little too loud.

"You do? What . . . I can't believe you even know what we do."

I told him about my friend Hiram Perez, who used the archives at Vassar College to dig through the hidden queerness at the formerly all-women's university, about how I love archives as a writer and reader because they let us inside the lives of those who came before us.

"Where do you work in the archives, sir?" I asked.

"Funny story," he responded, then paused. I was living out of a suitcase because I was visiting town. He was living out of one, staying with a friend, because it was his last weekend living in DC.

"I'm heading home, back to my mom." He grew up in Chicago, and had taken a job as the lead digital archivist in the Barack Obama library.

"That's . . . kind of a big deal," I said.

Jesse had long ago reclaimed his drink, and now he touched the back of my arm. I turned back to talk to him, my friend. No. Have some guts. I learned that if you want to talk to someone again, you should get their contact information before any circumstances can end your time together, goodbyes being the hardest moment to ask for a number.

So I breathed in quick and asked him, "Can I . . . just add you on Facebook?" He unlocked his phone, I typed in my name, and—like that—a notification bounced into my own.

I had a boyfriend, Wesley, back home in New York; it wasn't about a hookup. I liked Steven, and I wanted to keep laughing with him about the funny things you can find in the closets of history.

From Number 9, we moved on to Cobalt, where my friends Denne Michelle and Douglas joined Jesse and me for the promise of a dance floor whose surface was so sticky it took effort to lift one's feet from it to move. Douglas and Denne Michelle couldn't stand it, but Jesse and I didn't mind, walking around and dancing, Jesse knowing everyone. I stood against the wall, or chatted to him, sipping gin drinks. Steven did come, and we did dance, my arm in the small of his back. He went off again with his friends. I went back and found Jesse. I could see Steven across the dance floor until I couldn't. I smiled, remembering the notification—waiting for me already—that I'd managed to send from his phone to mine.

FROM MY ARCHIVES

2020.2020.journ.010, Friday April 10, 2020 (Good Friday, Journal Entry)

No samples. They're not coming. We didn't do anything. The sirens, they don't stop.

2020.2020.email.001, Monday, May 4, 2020 (Email Inbox)

The New Yorker: Seattle's Leaders Let Scientists Take the Lead. New York's Did Not.

2020.2020.email.002, Wednesday, May 20, 2020 (Email Inbox)

The New York Times: Lockdown Delays Cost at Least 36,000 Lives, Data Show

Lede: Even small differences in timing would have prevented the worst exponential growth, which by April had subsumed New York City, New Orleans and other major cities, researchers found.

2020.2020.journ.0011, Thursday, June 25, 2020 (Journal Entry)

It's Thursday, and Ngofeen and I cook together because we cook together every Thursday. I chop onions. He pounds ginger and garlic. We're talking about the RNA sequencing story, all the audio he has.

"Amore, I don't know what to do with the story. I don't know how to tell it and how it ends."

"Because it ends with failure," I laugh.

"Don't say that, it doesn't."

"No, I mean, it does. It's not our fault. We tried. But in the end, we never got the samples, we never sequenced them, the city didn't shut down early enough." I pause. "People died who didn't need to die."

"Maybe failure is the end of the story," I say. "Because we did everything we could. I wouldn't do anything different. I don't regret the work we did. But it didn't change anything."

He's silent.

"Maybe," he says.

He pounds garlic and ginger. I chop onions. In his kitchen, then, nothing but the smells that electrified our noses.

....

2020.2020.trans.004, Monday, November 9, 2020 (Zoom Call with Steven D. Booth, continued)

JSO: So we became friends because when you told me you were an archivist, I geeked out, and you geeked out at that. So just tell me about how one becomes an archivist and why you decided down this career path.

SDB: OK, so my name is Steven, last name Booth: B-O-O-T-H. I have been an archivist for about 13 years now. I currently work for the National Archives and Records Administration, where I have been since 2009, and in my current position I manage the audiovisual collection for the Barack Obama Presidential Library. Let's see, outside of my 9 to 5, I'm involved in professional organizations such as Society of American Archivists, also cofounder of the Blackivists Collective, which is a group of Chicago-based, trained Black archivists who work with individuals, community groups, and local organizations to help them document and preserve their legacy and history.

I became familiar with the profession during my senior year at Morehouse College in Atlanta, where I was studying music. And my music professor suggested that I go to a library school because I had really good research skills, especially as it related to music history and music theory. I ended up doing an internship at my campus library and I visited the archives one day.

And it was in that experience of them bringing out these historic materials and really valuable letters and documents and notebooks and diaries that I thought like, "Oh, wow, you know, we—and by we I mean Black people—we really do have a rich, unique history that's not always captured in books. And so I thought, "How cool will it be to kind of unearth those stories?" And so I actually went into archives with the expectation of becoming like a writer or researcher.

I ended up going to Simmons College, where I studied with the first African American archival educator. In my second year of grad school, I started an internship at Boston University, where they were processing a collection of Dr. Martin Luther King Jr. That internship turned into my first professional job.

JSO: And what exactly is an archive? I feel like so many things might almost apply that I'm not sure how to narrowly or specifically define it.

SDB: *laughter* You have kind of like three versions or three definitions for what an archive is. So archives can be materials of historic or enduring value about people, communities, and places and events. So those materials can be like letters, email, your diary, a notebook, text message, a photograph, a video. They can be analog or digital. And then you have institutional archives that are dedicated to preserving documentary materials, like the National Archives or Records Administrations. And the third can be a physical or online space where these documents are preserved.

It's important for people to understand that

archives are real things in real places. Even digital or online archives are kept in a physical place. There's this notion that the cloud is really just the cloud. No. It's a digital thing on a server which sits in a facility. Even your Instagram is sitting some place, or many places, and anyway it isn't the same thing as an archive, because you can't be sure whether the technology will still be available in 10 years, and if it's gone, we lose that content. Remember MySpace? Archives preserve materials for generations.

JSO: How do materials usually get into an archive, and who has access to them once they're there?

SDB: Well, it depends on the institution and the repository. I think over time, we've seen archives as only accessible to the "serious scholar and researcher." It's interesting to see the shift happen where we're starting to recognize that not only do we need to diversify our collections, but we also need to diversify our user communities.

JSO: I've always thought of archives as like catching lost histories. I might not have learned very much about ACT UP, for example, without archival footage that was used in *How to Survive a Plague* and *United in Anger*. But some of the writing you sent me made me think deeper about the fact that if you don't get captured in an archive, your story is then truly lost forever. And we know there are biases in whose lives end up archived, too.

SDB: I think that's where the beauty of community archives come in. That's a big conversation within the profession now, a model where the collections and materials don't necessarily have to reside in a traditional mainstream repository. Collections and materials can live, thrive, and be preserved in the community that created them, whether that be an individual, a community—a queer community, a Black community—or an organization.

JSO: I've been thinking a lot about how archives can seem extractive. They take documents and information and stories from a community and place it in a serious institution where serious scholars can come and "discover" serious history.

And I saw you on a panel, what, two weeks ago? It feels like forever, there's been a lot of shit happening, but one thing that came up in the panel was archives through the lens of a feminist ethics of care.

SDB: There's a connection to it, especially from a professional standpoint, as it relates to your relationship with the user, as it relates to your relationship with the records, and the subject material. Archivists are not impartial.

I'll give you this prime example. At the Center for Black Music Research, I was put on the reference desk and received a request from a woman who claimed that her father was a prominent jazz musician. And she wanted to verify this information in order to share with her grandchildren. She had heard tales, you know, that have been passed down in the family.

So she gave me the name and I did some digging. Come to find out it was actually true. He had played with Duke Ellington and some other really great musicians. And his name was in a couple of discographies. I shared that information with her. I made a photocopy of it and sent it to her. And, you know, the conversation that we had on the phone, she was overcome with emotion. I cried too. I think this is also the beauty of the archives. You see yourself in the archives. But not only that, you may need someone being truly willing and able to assist you in that process. In this validation that the story is true and now I can share it with my children, then they can know this piece of their history and heritage. It's easy for archivists just to kind of see our work as day-to-day work.

I think materials, objects always evoke feelings and emotions. I think it's important to really be cognizant of that. I think in a lot of ways, at least for me, archival work is rooted in a form of spiritual practice.

....

2020.2020.journ.012, Monday, March 9, 2020 (Journal Entry)

The iPhone in my hand was on speaker. I held it up to my mouth close enough to speak, but I wasn't speaking. Silence on both ends. The workday had been long, and—for 2 hours already—I'd been holding this phone in my hand speaking and listening in turn.

I hate talking on the phone. Even sometimes with these closest people to me, at the end of a long day, I can't even pick up if they call, except if they call three times in a row.

It's March 9, the city is doing nothing, and everyone believes this crisis is being overhyped by the media for clicks. To do something, anything, I start messaging my friends in nightlife and those who go to sex parties. We need to have the conversations that our elected leaders are not. We—queer people—need to know the risks of our events and gatherings, that they might harm us or those close to us. We need to lead the city and nation in acting fast to care for one another.

"What did you say?" from the other end of my phone.

This is the third call today with the third different sex party promoters and hosts. There's nothing more dangerous, if COVID-19 is in your community, than a sex party, but not because of the sex. That's the safest thing there. What's risky is the air, the spit, how many bodies fit into such a small space, how close mouths are when two people are fucking, how hard we breathe out, how desperately we inhale.

On these calls, I'm imploring folks to shut down before the city makes them. It's the right choice. It could save lives. COVID-19 is already here in our city.

But I have a second mission, too. I care about this community, I go to these parties, I know that lost income from sex work or hosting sex spaces won't be easily replaceable by unemployment. I need to tell these people that they need to plan for weeks, months, and maybe more than a year where the city or state will make their parties illegal.

"When the city does shut you down . . ."

"No, after that. How long did you say?"

"Months. A year. We don't know. It easily could be."

"Oh come on, that's how I know you're buying into the panic you see on TV."

"I'm just speaking based on the things we know now and the data I see, I'm just telling you to prepare."

Silence again.

"Oh come on. There's no way." But the silence between sentences speaks as much as these words. Even if he thinks I'm bullshitting, he's heard what I think.

"I told you he was full of shit. There's no way we're listening to him and canceling this Saturday," a voice from far away. I knew I was talking to a room full of fags, my people, the one who hosted the party and all the employees who got paid to run it. I was fucking with their income, especially the people working there whose income that night might have been what they needed to eat that week.

Mostly, I remember the silence, the quiet. This could be months or years, I said. Silence, an extra beat. I hung up the phone, and my arms fell down at my sides, heavier than I knew them to be.

....

2020.2020.journ.013, Friday, June 26, 2020 (Journal Entry)

The report that James put together on PrEP for COVID-19 was incredible; its depth and nuance earned us a meeting with the people who could do something about how slowly things seemed to be moving. Peter Staley emailed the report to Anthony Fauci—his friend from the 1980s and 1990s working on HIV science advocacy. If Peter sends Tony something, he'll at least look at it. The relationships that Peter and David and Mark have from their decades of activism, and Wafaa's connections from her decades of public health work and research, mean that sometimes, when we have an idea, we can try, at least, to do something about it.

What am I doing here?

Call with NCATS today, James and I are leading. On a phone call earlier this month about James's report, we asked Tony Fauci and Francis Collins about high-throughput screening for new COVID-19 drugs, they said they don't know much about where we're at. But the guys at NCATS will! Do we want to set up a call with them?

We do. That phone call is today.

National Center for Advancing Translational Sciences, or

NCATS, is a center within NIH that does public work turning research into therapies that could get into patients.

It's a few of us from the COVID Working Group and four scientists from NCATS in a grid on Zoom. We ask about which cell lines are being used for high-throughput screening. We ask which libraries are being used, and why. We ask who's doing the work and why. We ask about data openness. Matt Rose—a Working Group member turned friend—described these guys after the call to me as the nerds in the basement, and he meant it as a good thing. These scientists were out there trying to find a new drug, already FDA approved for something else, that would work for treating or preventing SARS-CoV-2.

They're looking for drugs against SARS-CoV-2, but they can't grow the virus in the lab.

Here's where I have to back up a bit. SARS-CoV-2 is a virus, a deadly one that transmits easily. To work with it safely, you have to have a special kind of lab. A lab that prevents the risk of transmission from the cells you're working on to your own cells. This is dangerous to the individual scientist, of course, but also to their community, as they could pass their SARS-CoV-2 infection on.

The special labs needed to study SARS-CoV-2 are called biosafety level 3 labs. I've never worked in one; my labs all were biosafety level 2; we could grow human cells (which I've done) and infect them with an infectious virus (which I've done), but not an infectious virus that can infect humans. I was working then on MMLV, a mouse virus similar to HIV but that I could work on easily precisely because it could infect mouse cells but not my cells.

NCATS is testing for new SARS-CoV-2 drugs, and they're using robots to test thousands of drugs at once, but they can't grow SARS-CoV-2 in the lab because they don't have a BSL-3 facility. Building one would cost $10 million. We're four months into a global pandemic that has cost our nation alone trillions. My boyfriend lost his job. And the agency that's trying to find a molecule to save lives isn't working as fast or as well as it can because they didn't get a $10 million building built as fast as it fucking could have been built.

I am shaking with rage. I'm all out of shock.

I end the call with my rant, and I've gotten good at it by now. The entire governmental response to this virus starts and ends with a vaccine. And we all hope one comes. But if it doesn't, we need drugs, and good ones. And look at HIV. No drugs worked for HIV until we designed drugs specifically *for* HIV. Yes, for COVID-19 that could take years. But we don't know how well or how long our immune system will protect us from reinfection. What if it's like flu or other coronaviruses, I ask.

This thing could be here for years, for decades. I'm not saying it will happen. I'm saying it could. We need to start basic research into how this virus works, the kind that was necessary for good HIV drugs, and hep C drugs, and we need to start it now. If we wait until the vaccine fails, it will have been too late.

Rant over. Heads nod. James and Matt and I all text-message about how batshit it is that the government won't give the nerds in the basement the small thing they need to do their work. Peter texted Tony after the call; exhausted, Tony promised to "add it to the pile."

"It's like they're actively trying to kill people with neglect," I text Devon. I think about teaching in the fall. I run in the park and use my weight bands on the roof. My nerves are crawling up the back of my throat even as the sun warms my shoulders. I wonder if we're gonna make it through this. I think about Mark and David and Garance and Wafaa—she was an infectious disease doctor in Harlem in the 1990s—on these calls. I wonder about what marks it will leave behind if we do.

2020.journ.014, Monday, July 27, 2020 (Journal Entry)

It's Monday, and she's tired.

I got invited to participate in a 9 a.m. call this morning that I promptly did not add to my calendar and so missed. I woke up at 10, the moment the call ended. I cut myself cooking twice last night and so I have two fingers sore and wrapped in taupe plastic.

"Shit, I think I missed a call," I told Devon lying in bed. He smacked my shoulder lightly.

"I told you to keep a calendar."

"I do!"

"Then why are you always missing calls?"

"I have a lot of calls."

He kissed my forehead.

At 4 p.m., I joined a Zoom call with Steve Holland from the NIH. He works for Tony Fauci and we're talking to him about why in the flying fuck the NCATS folks can't get access to a BSL-3 lab to grow SARS virus.

Steve is the person who could make it happen. The NIH has BSL-3 labs where this work could get started tomorrow, but the folks who normally work in those labs are protective of that space. They don't wanna give it up.

Steve acknowledges that he hasn't thought of it. Why exactly do they, at NCATS, need to grow this dangerous virus he asks.

Well, as a molecular microbiologist, I can take this one. James has been thinking about it nonstop too. We pass our arguments back and forth. They need to test different cell lines. Right now, they're subcontracting these experiments and so can't troubleshoot.

"Isn't the pharma industry doing this?" he asks.

We don't know, because industry isn't required to tell us anything. Plus, every company has their own set of drugs to test, and they won't even tell us what those drugs are. We need an open process, we say, with every molecule we have, with protocols immediately available, and follow-up open to anyone. No company would ever do this.

"It will inform," I say, "rational drug design." That's when scientists build a drug to block a particular protein instead of testing everything all at once. But knowing what type of molecule works well helps narrow that search or fine-tune early molecules.

When we get off the call, Peter says that he thinks NCATS will get the lab they need within a week or two. When we end the call, I feel like we might have done something. We might get the nerds their lab. It's unlikely, but possible, that a molecule they find will be a potent SARS-CoV-2 drug. Who knows. For now all we can do is all we can do.

For now, my foot is hanging out in our dog Max's bed, and he's asleep on me. For now, my fingers are held tight in their plastic bandages. The pain has abated and the tightness around my flesh feels almost calming. A reminder that when something does cut into our flesh, we can help it heal.

2020.2020.journ.015, Thursday, August 13, 2020 (Journal Entry)

I'll keep it quick today. James messaged me on Facebook around 11 a.m. The last few weeks, I've been sleeping in late, only getting up to walk the dog at 10 a.m. or even after. I reason that, in three weeks' time, I'll be back to teaching, and I'll want and need to be rested when all that kicks off.

He messages me around 11 a.m., and I'm having my second cup of coffee and deciding what to work on today. Teaching stuff? Writing stuff? A mix of both, I decide, and so, this morning, I'm trying to finish an essay draft on viral evolution, human evolution, the inevitability of change, and how we can accept, and even use, life's changes instead of being at an endless "war" with viruses, bacteria, aging, death, a war we can only ever lose.

"We won on the HTS thing," James texts. HTS is high-throughput screening. "Screening is estimated to start in early September at BSL-3." They freed up some preexisting lab space to start ASAP and are upgrading the current NCATS facility to be a BSL-3 lab to be able to screen against this and other pathogens that might come in the future.

It's been just over two weeks since our last call, after which Peter said they'd have the lab within two weeks. Without our work, James's work of identifying screening as a necessary thing that likely isn't happening, and then all our work together across weeks to figure out what was happening, what wasn't, and why, without all that work, no one in power at the NIH would even know, now, that screening wasn't happening.

We did something. Who knows if it will matter. But we found something that was wrong and, at least for this one thing, we put in the work and we fixed it.

"!!!!!!!!" I respond. "Too early for champs?"

On a day like today, it isn't. But I didn't indulge until our 5:30 p.m. group call. I'm drinking champs, Peter red wine, James white; Wafaa is in her office, without a beverage in her hand.

"It's a cool victory," said Peter on that call. "Especially if they find a molecule." The NCATS screening could identify a new drug that works against coronaviruses like SARS-CoV-2. If that happens, then our work would have done something that might save countless lives.

I think it's a cool victory regardless. My years doing science taught me that you can't hope for the perfect outcome. You just have to do what needs to be done and have some faith. If you don't look for a drug, you'll never find one. If you don't try to sequence the RNA, you won't sequence it. If you look hard, you still might not find a drug, but that doesn't mean the looking wasn't worth it.

The looking is always worth it. And starting in just two short weeks, the NIH will be looking.

....

2007.2007.trans.001 (From the ACT UP Oral History Project)

Sarah Schulman: OK. All right, let's go back to the FDA demonstration. What did you want to say about that?

David Barr: Oh, OK. Well that's a good story. Well, I don't know if, have people spoken about the FDA demonstration?

SS: Everyone has a different story.

DB: Oh. Well, all right. Then I have my story.

SS: OK.

DB: I have to tell the story this way; I can't tell it any other way.

SS: Go for it.

DB: All right? All of this work was about a lot of people. Right? You're interviewing me, so I'm going to tend to focus on my work.

SS: That's what everybody does.

DB: I just don't want to make—I just want to acknowledge that this was about a lot, a lot of people.

....

2020.2020.expo.001

I watched the FDA action—Peter and David and Garance and all the others—in *How to Survive a Plague*. The film, by David France, is not without its critiques, including by Sarah Schulman, who was in

ACT UP (which France covered as a journalist). Schulman and Jim Hubbard released a film of their own, *United in Anger*, the same year that France's film came out. Jim and David were roommates in 1987. Jim was an experimental filmmaker and shot some of the original film used in both movies; the rest was filmed by various people making records for archiving the movement in real time. "Let's film this. This is our lives," Hubbard said in 2013—in a discussion with France—of the 1980s in ACT UP. "There's always a political aspect to aesthetic decisions," he added, about turning that footage into a film. France's film, he says, is about a thin slice of ACT UP, individuals like Peter and Mark and David, who worked on drug access and development. These men were all white and cis and gay and came from educational privilege. Jim's film was a portrait of a movement, of ACT UP, built out of the ACT UP Oral History Project.

In 2021, Schulman published an authoritative history on ACT UP, *Let The Record Show*, a book based on the oral history archive she helped build. She wrote the book to reflect the horizontal nature of the organization, with lots of individuals working in countless groups on various, at times overlapping, projects.

The "heroic individuals" myth that Schulman identifies in David France's film, "aside from being inaccurate, could mislead contemporary activists away from the fact that—in America—political progress is won by coalitions." I wish I'd seen *United in Anger* first, but—even as a queer scientist—that film didn't reach me until much later. I'm grateful that, finally, it did, and alongside it the ACT UP Oral History Project, and after that Schulman's book itself.

I'm writing about NCATS and RNA sequencing not because it's the most important work we did in the COVID-19 Working Group. It's just some work that I did, two stories that feel mine enough that I can tell them. I think the most important successes we've had so far are Charles King's work with Housing Works to push the city to open up hotel rooms for people—including homeless people—to isolate in with COVID-19 or to have a safe space to prevent getting it. In 2021, the city would undo this decision. How temporary our victories can feel. Housing Works also pushed the city to adopt harm reduction policies and not zero-tolerance spaces for drug and alcohol use to get people into these rooms and to help them stay as long as they need.

C. Virginia Fields and her coalition, National Black Leadership Commission on Health, led work to get New York to disaggregate COVID-19 cases and deaths based not just on zip code but on race. This change showed unequivocally and for the first time who was getting sick and dying of the pandemic in our city and country: Black and Brown New Yorkers.

With Charles King and Wafaa El-Sadr, we worked to get the city's test and trace to be transparent and run by public health officials. Learning from Wafaa has been one of the biggest gifts this crisis has given me; she's a doctor and epidemiologist and she knows . . . everything. And she cares about public health through a social justice lens. So far, we lost the test and trace argument with the city and are still losing, but we keep pressing, every week. Our CWG members who work at NYCLU helped to pass state legislation ensuring that cops and ICE officials couldn't access COVID-19 test and trace data; Cuomo refused to sign the bill for over six months.

I don't know. Most of what you do fails. But if you didn't do it, you'd have no chance of succeeding at all. You throw shit at the wall, and see what works, and plan, and then try to throw more and better shit. Doing this work has hopefully saved some lives. In a way it was selfish. I needed to do it. In the face of an impossible world in the midst of a deadly pandemic, I'm so lucky I felt like I could do something. It kept me sane. The struggle, if it saved no other lives, was a part of how I saved my own.

....

2020.2020.trans.005, Monday, November 9, 2020, 8 p.m. (Zoom Call with Steven D. Booth, continued)

Joseph Shannon Osmundson: You know, I came to the archives for this project from two films that used archival footage, one called *United in Anger.* The other is called *How to Survive a Plague.* They both use a lot of the same archival footage.

I love both of these movies, I should say. But one of these films is almost only white, and the other film is less white but still pretty white. But the archives of ACT UP have all sorts of stories in them. And I wonder about whose stories are the ones we choose to tell.

Steven D. Booth: There is this push within archives and libraries to diversify our collections and I think that's great, I think it's beautiful. I think it needs to be done. A good example is the director of the Alabama Department of Archives and History. They recently posted this press release acknowledging the White Supremacy that exists in archives. The former director was very intentional about not collecting anything related to the Black experience in the state of Alabama and how they're going to actively work towards doing that better now. But of course, that's not enough.

And even then, I'm not sure if those who are doing cultural *production* are necessarily aware or are doing that same work. So the problem has to be fixed on both sides. Archivists can be good at talking to other archivists, but we need to be better at talking to other people.

JSO: Are archives antiquated in the world where Google archives everything we type into a document or email or message?

SDB: No, I don't think so at all. People are still archiving their Google email with institutions in order to have it accessible to a community of users at some point. I can't go to Google and say, "Give me Joe's emails from 2017!" They wouldn't do it. But if you deposited your stuff, we could. And their archives aren't for us, but to make money by selling you things. This really illustrates the public good that we do.

We're in this moment with COVID-19 and racial injustice, and I think people are thinking more about their memory and their legacy and working towards documenting their experience in order for it to be preserved in perpetuity. We've seen a rapid emergence of Black community archives around the country as well. We understand that people want information and they want access to that information, especially digitally, because that's the culture we live in.

I think at the core, that's what archives are: documentary evidence that something exists, that something happened. Just capturing conversations, transactions, whatever you can think of, you know, that's what it's doing, it's serving as evidence.

As Black archivists, we've worked in these institutions for so long that we know where the gaps are. And also having roots in the city, we know where those unique things that are very specific to the city of Chicago are, and the people who have those experiences. But you won't find many of those narratives and experiences in mainstream archives and I think that has a lot to do with how "notable" someone is, it has to do with an institution's collection development policy.

So I think it's important to put the responsibility of this work back into the community, to let them know that they can do this themselves. I think all collections are part of some community.

JSO: My first interactions with archives were learning about activist history or doing activist work now. Like, it was ACT UP's archives and then my friend Hiram Perez's work on finding queer stories in archives from the twentieth century. Do you see an inherent connection between activism and archives, either with activists writing their stories down or with community members and scholars unearthing activist stories we don't know well enough?

SDB: Well, I think it depends on the researcher, and I think everyone has a different introduction to archives. The act of collecting and organizing and cataloging and providing access to materials, there is nothing about it that feels like work to me. It's what brings me fulfillment. And maybe a form of activism is providing access to information and resources because I think information and resources really can empower communities to effect change in their own lives. And so maybe that's the activism part of it, but to me, it's just that's just my work, right? That's just what I've been put here on this earth to do. Like, you know, you'll never meet a better researcher. I can find anything and exhaust all possible options before I give up.

JSO: Oh, you're saying you're a Virgo.

DSB: Double Virgo Cancer. It's the perfect job for me.

JSO: The Cancer is where you get the emotional side from. Like when you give someone an artifact and cry with them about its importance.

So until now, most of your work has been on the archiving side of the project. Do you feel like at some point you're going to come back to what brought you to the archives in the first place, which is more of the storytelling side of it?

SDB: Yeah, and actually I feel that shift happening now more than ever before. So I'm really interested in Black queer narratives, particularly like as it relates to spaces. I'm thinking about clubs or bars or even house parties that existed. And I think my interest in that really stems from like having lived in DC and getting to DC when there was like a really booming gay scene. And then within my first two years of being there, a lot of establishments just closed down, which left us with, you know, just like a handful of places to go to.

JSO: And we met at a DC gay bar.

SDB: On 14th and P. So many more of those places that were literally on the brink of closing but were being saved because somebody was donating X amount of money every couple of months. But because of COVID-19 those places might no longer exist. I did a preliminary mapping project of some former Black queer spaces that existed in DC and so that has been an interest of mine to think about how to do that here in Chicago.

JSO: One last question: If you were making an archive, what's one thing you'd want to be remembered in 100 years' time?

SDB: If someone wants to read this 100 years from now, I would want them to know that, at this point in my life, as a man of a particular age . . .

JSO: You're thirty-six!

SDB: Oh, don't do that. Thirty-five. Thank you. I'm officially middle-aged. But what I would want them to know is that at this point in my life, I am starting to take a step outside of my comfort zone and really explore some new opportunities and possibilities as it relates to my life and the contributions I want to make to the world. And so I would hope that in 100 years, there's something to show for that. Whoever the reader of this is can pinpoint back to this moment and be like, oh, this was the start of X, Y, and Z that came later, you know. If that makes any sense.

JSO: That makes perfect sense. I will just speak on behalf of the world. Or just for myself. But our relationship started years ago. I can say that I'm grateful for your contributions to my life in the way in which you make me think harder and be better. I've learned so much from you!

And also just like you are fun and goofy and great to hang around and I love you so much. I'm grateful for that time. And I love you.

SDB: Anything for you.

....

2020.2020.expo.002

They always say don't meet your heroes. I met some of mine, and they aren't my heroes anymore. Alex Chee, through writing. Mark Harrington and David Barr and Peter Staley through COVID activism. They aren't my heroes. Heroes don't make you laugh, you don't worry for their health, you don't ask yourself whether they're happy. You don't see the insides of a hero's life. They aren't, they can't be, human. You don't disagree with them about a tweet but keep on loving them just the same.

David Barr wrote me, after reading all this. **July 30, 2021 (2021.2021.email.001)**:

> I wrote a piece a bunch of years ago dissecting the whole construct of PWA (person living with AIDS) as hero. We created the "hero" construct as a way to change the ways in which people with AIDS were presented at the time—we were either victims or vectors. The whole notion of PWAs as heroes guided not only our individual behaviors, but the communities' behavior, particularly among activists. The PWA was the organizing principle around which everything revolved. We were heroes. But we died. There was no time for grief. The work could continue in the name of the fallen hero. The PWA and the idea of PWA empowerment gave everyone strength, regardless of serostatus. But, of course, it was a myth. We weren't heroes. And we were scared to death. The construct was useful to survival but had long-term harmful effects. We weren't allowed weakness or grief or fear. Anger was the only emotion (besides joy) that was

permitted in ACT UP and it fucked many of us up very badly later, myself included. A lot of these feelings came back up for me as COVID blew up. James was living here in my house, frantic and starting his work. I was really reluctant to get involved, though I did eventually and am glad I did. I don't know how helpful the work has been, but some of it has been helpful and, as you say, it was personally helpful in that it gives you something to do, some way to respond and not feel so helpless. In that sense, it's selfish, but in a good way.

2021.2021.expo.002

I'm sorry, David. You were a hero to me once. And I needed you to be; I needed to see people queer like me as strong and fearless and resilient and angry. But that made you less than human. You weren't fearless. And I'm sorry.

2020.2020.expo.003

I love the way David says "Hey Petey" to Peter every time we meet on Zoom, a cutesy voice singsonging back and forth between them, a pet name they've used since the 1980s. I love calling Alex Chee and comparing the dinners we're preparing for ourselves, trading ideas for cocktails. Together we named a cocktail of my own invention—a gin gimlet topped with Prosecco—a Roast Beef Tivoli (Roast Beef as the French pejorative for English people, the gimlet, and Tivoli Fountain for the addition of the Italian wine).

Heroes aren't people. And if all your stories of change being made relies on heroes, you don't imagine you can make change for yourself. And we can. And we must. The people who fought against inaction in the HIV/AIDS plague years, including Alex and David and Peter and Mark, are just people, they're fucked up and human and messy and everything else we all are.

The archives are a way of getting to know. Archives are not a well-curated story, not yet. Archives contain information, all the messy information of a whole person's life. Their diary, their email, all the photos on their phone.

The archives are a way of getting to know people. I don't know what the archives will show, what my work will be worth, what meaning a reader will make from the archives I've shared here. I'm grateful for these people I've gotten to know. Many of them I've only ever seen on screens, on Zoom. A friend who knew Wafaa before told me, just this week, that her signature look is cowboy boots, she always wears cowboy boots. I've spent hours with her, but have never seen anything but what you see on Zoom. I know her only from the shoulders up. This woman I feel I know so well, now, who I still have never seen. I don't know what to make of it. I feel so close to all of them. All of them I something-like love.

Matt Rose pings me a few times a week, "And how are you doing?" I ping Matt back, "She's alive." A few times a week, I ping Matt, "You hanging in?" and he pretty much always is. We didn't make a change with the hours we spent on RNA sequencing advocacy. We did on screening at NCATS. Who knows if it will matter in the end. I'm so grateful that I could do something, anything at all. The love I have for queer people, the moment of brief terror I shared on the phone with someone who hosts sex parties for a living, who didn't yet know that this virus was coming, coming for us all. I was just trying to love. I love Steven. I am so grateful for all I've learned from him. Activism is spiritual work, archival work too, and friendship, maybe, most of all. From community—from friendship—the best activism and the best archives come. The love of trying and failing but trying again. Trying because the only thing more horrible is not trying at all. That love, and how we tried to live it, more than anything, is what I want the record to show.

10

On Endings

Do Plagues Ever End?

When I was 15, I had a recurring nightmare of sticking my finger into the coin slot of a pop machine. My parents usually let me get a store-brand pop for a quarter outside the Safeway down on Olympic out where the Jensen farm used to be. In the dream, I put in a dollar, grabbed my pop (Shasta Cola, I believe), and my greedy fingers reaching for the three quarters in change met the coins, yes, but also were punctured by a needle.

Rumors of HIV-infected needles left in coin return slots of Coke machines and pay phones—remember those? These rumors circulated online and off from the mid to late 1990s, with newspapers constantly debunking them. But the debunking only seemed to validate the rumor. Why would it be on the news if it had really never happened? In 1999, in Pulaski, Virginia, two people were stuck by needles when they reached into a coin slot of a pay phone. A prankster, officials figured. The needles were new and never used, some kid trying to retroactively make the rumors true.

No one was infected with HIV or anything else due to these needles.

At 15, I kept dreaming it true.

I was afraid of kissing girls because of the virus. I knew what I knew. Kissing couldn't spread it. Fear never listened to logic, especially not at 15.

We live with every virus on earth, whether or not they are in us yet. We live with every virus that could be in us—infect us—in a different way: they are a proposal, a knock on the door of our cells, a molecular desire for entry, for two things—one virus and one cell—to become one.

At 35, I started taking HIV drugs—for good this time—after a boy I had sex with took off the condom in the middle of sex. He was on PrEP. He felt it was safe. He didn't ask my consent; he knew I wouldn't have. I didn't stop because it would be awkward and I didn't stop because I liked the boy; I didn't stop because we were having fun. Years before this, in this exact situation, I'd stopped. Now, I didn't stop because I knew I could manage my risk the next day.

The next day, I got on PEP—post-exposure prophylaxis—Truvada plus raltegravir. Truvada is two drugs, Emtricitabine and Tenofovir, that both block the virus's machines that copy itself. Raltegravir inhibits the viral integrase, the enzyme that sticks its viral genes into our DNA. One month later I continued on PrEP—pre-exposure prophylaxis—Truvada alone. I was born in 1983, the year HIV was first identified as the likely virus that causes AIDS. HIV in me, then, at 35, like at 22—when I was a false positive—like at 15, like at birth. Between 35, when I started PrEP, and now, the worry of HIV has receded; PrEP was a part of that, and so was the shifting culture around raw sex, gay sex, pleasure, and shame. Biomedicine provided PrEP; activism made it more widely available; queer community offered a pathway toward an exit to my lifetime of worry. HIV is still here; queerness changed its meaning.

I stopped taking PrEP when I stopped seeing strangers, in late February, in 2020.

....

The binary of illness/wellness is always porous, whether or not we notice. In *Illness as Metaphor*, Susan Sontag wrote, "Everyone who is born holds dual citizenship, in the kingdom of the well and in the kingdom of the sick."

I used to believe Sontag uncritically. Now I know that the two kingdoms don't exist; it's one kingdom, in a quantum state. We are all constantly both sick and well, sick/well, sick and well both, our body and immune system in conversation with a world rich in microbes, viruses, and bacteria, almost all of them either benign or beneficial, countless microbes in me here as I write and in you there as you read.

The quantum state of sick/well. Let me take cancer, that disease so long synonymous with death. Cancer is not a binary. We're willing to acknowledge this, but only when one has already crossed over into the sick category. Cancer has stages. Stage 1 cancer is still small, mostly easily treatable; stage 4 cancer is aggressive, malignant, *deadly*. Many prostate cancers require no intervention at all. They are subclinical. We live with them until something else kills us.

Cancer is not black and white but gray, and further, there is no white to begin with. There is no absence of cancer as long as we have a body. We make cancerous cells every day our cells divide, which is to say every day. In order to keep living, and keep making—for example—new skin to push the outside world out, our cells have to divide. With each cell division, a cell will make mutations. Mutations are the raw material of cancer. Carcinogens and sunlight cause cancer because they cause mutations and damage to our DNA. But without cell division, no life. To risk cancer, to make it day by day, is to live. The absence of cancer is death.

So cancerous cells are a normal part of life. In the popular imagination, we see the immune system as mainly protecting us from infectious disease, from bacteria and viruses. But just as importantly, our immune system protects us against cancer, the cells in our own body that mutate to become other, to pose a threat. At almost 40, I've certainly had cancerous cells in me, cells that mutate so they can divide and divide and divide. It's my immune system that finds those cells and kills them so that I, a whole organism, can survive. But because I don't have cancer doesn't mean I haven't had a cancer cell in me. I have, and survived it, so far.

This is why HIV was first called a gay cancer, Kaposi's sarcoma (KS) skin lesions one of the most visibly obvious signs of advanced disease. HIV kills cells of the immune system—CD4+ T cells to be exact—and the immune system can't find the cancer anymore, and so the cancers we'd survived for decades become something more deadly, something more visible, a red patch on our skin.

We are all at risk. We are all sick/well. KS might be in me now, kissed up on and held down by my T cells.

Viral infections, too, are a sick/well. With HIV, HIV-positive individuals can be undetectable, no virus in the blood, no risk for transmitting HIV, a projected life span the same as everyone else's. And with herpes or chickenpox, once we have an initial infection, the virus is in us always, but dormant. Are we infected then? Yes. But are we sick? Do we have symptoms? No. We are sick/well. If our immune system were—flash—gone, the viruses would all come roaring back.

Our immune system is constantly talking to our viruses and the cancers, the viruses constantly present and able to reactivate (cold sores, shingles), but mostly just there, not just in us, but *us*. Virus/me, sick/well, cancer/me, always.

One's relationship to sick/well often has more to do with one's material circumstances than the circumstances of one's body.

Those living with HIV who are white and wealthy and in New York City may have a very different experience (well) than those living with HIV in the rural South without a healthcare facility anywhere nearby (sick). HIV is material, but it's not the only thing that is. And it's not HIV that determines our health. With so many diseases, the fact of the disease itself is often less important than who one is and how one can access care.

But acute viruses, like COVID-19 and influenza, come and go. Here the binary of infected/uninfected holds up to stricter scrutiny.

But even those uninfected live with the virus—living is an active verb—it lives with us, too.

When I wash my groceries with ethanol after coming home from the store, I'm living with COVID-19. When—in June—data on fomite transmission implied that the risk from these types of surfaces isn't so high, that the air in the grocery store is much more dangerous than the groceries themselves, I decide to stop washing everything that comes into my home in ethanol. I'm living with COVID-19 still.

When I wear a mask. When I stay home. When I don't stay home and knowingly or unknowingly accept whatever risk there may come, I'm living with. When I get a vaccine, my body is reacting not to a virus but to a part of one, building protection against disease. When I get boosted. I'm living with, a part of the virus is living in me, its memory hopefully long-lasting, the echo of a lost love.

I don't mean to imply that there are no material differences between the HIV-negative and HIV-positive categories. Or COVID-19-negative and -positive, either. Of course there are. And being HIV- or COVID-19-positive puts one at risk of dying from that virus or from how people perceive it. HIV is a material thing, its genetic material living in HIV-positive people.

Being *infected* with COVID-19 is different from *living with* the virus. And living with does not mean we must let 'er rip, let the virus move through our communities unabated. No. Living with COVID-19 means not denying its risk, but trying to minimize the harm for myself and all the humans I care about, which, on my best days, is all the humans there are.

....

Look more closely at Kaposi's sarcoma. KS is a cancer. Cancer comes from our own cells. Our individual cells—except stem cells—will die after they divide a certain number of times. Cells that don't die, but learn how to divide and divide and divide are cancerous: they can eventually multiply so much that they kill the rest of us.

Imagine my arm with a blood-purple KS lesion. Wash your hands, glove up, and scrape off some of my skin cells. Move them to a tube with saline and formalin, salt and preservative. Chuck it on the microscope next to some healthy skin cells.

What do you see? Hallmarks of cancer cells: different colors, odd shapes, overgrowing one another.

Look closer now, beyond what a light microscope can see. Chuck it on an electron microscope.

What do you see?

A virus.

Kaposi's sarcoma is caused by a virus, HHV-8, human gammaherpesvirus 8, or Kaposi's sarcoma-associated herpesvirus, an enveloped double-stranded DNA virus from the family Herpesviridae. Once this virus infects a cell, its gene expression instructs the cell to grow and to divide, to become—in essence—a cancer. The virus doesn't copy itself just by killing cells and making more virus, but by instructing the cells it lives in to copy *themselves*, and the virus along with them.

HHV-8 infects millions of people worldwide. The virus is spread in ways we don't yet understand: through spit, we think, and sex, and blood, any way bits of our fluid might meet another human being. Millions infected with the virus. Very, very few cases of KS. So few—in the United States and the United Kingdom—as to be barely measurable.

Many viral sequences choose or take this path. They pass themselves along with the cell they meet and infect. Most of these viruses don't even cause a cold. Some viruses are already in us, unable to escape, the memory of something we already survived. Retroviruses, like HIV, that infected us millions of years ago have lost the ability to leave our cells. Endogenous retroviruses we call them; viruses that are inside and cannot leave. Their genome—their DNA—is in us still, the molecular memory, the DNA sequence, of their replication. We can't take them out; they are us. They're viruses we survived and named as our own.

Endogenous retroviruses, from the family of viruses that includes HIV, make up 10 percent of all our DNA. These viruses can turn themselves on, and do, but can't leave us to go infect another. They can only be passed down *with* us. HERF-T and HERV-W; HER4-like and HFV; HML6 and HML9. Can you feel them in you now? Or do you feel just like yourself? Regardless, there they are. There they will remain. One day, I dream of HIV this way: inside us, doing no harm we can feel.

First do no harm. Sometimes viruses do the opposite of harm. My favorite story of them all? Herpes simplex virus 1 (HSV-1). Most of us have this virus in us, too, at least one of them, and have since before we remember. Yes, HSV-1 causes cold sores, and I hate cold sores. The way they sting before you even notice a pimple; the way you pray it's just a pimple and not a cold sore when you know good and well it already has that cold-sore sting; the way it weeps clear liquid; the way it takes off, takes over, killing cells and sloughing them off and making an open sore as your body tries as it might to heal.

But most of the time, most of our days, this virus is just in us, not making cold sores, and on those days, we ought to be grateful for it.

Research just about a decade old implies just how long we've lived alongside herpesviruses like this one. In fact, if you try to grow up mice in cages with no viruses or bacteria at all, they never develop an immune system.

In mice, infection with a herpesvirus actually protects against infection with a dangerous bacterium that can cause food poisoning. Herpesvirus infections can also help the immune system respond to malaria or to other viral infections. The hypothesis is that herpesviruses, as common as they are, as persistent as their infection in us is, have evolved to essentially be a component of our own immune system. The virus is always present, trying with its viral might to become active again; our immune system is always there, trying to keep it quiet. This ongoing, lifelong conversation primes our immune system to respond to another pathogen. The virus has been with us so long that without it, our immune system may be under-responding.

Viruses have been here all along, an invisible part of who we are as a species. Our bodies evolved with them. Like cancer, the only way to rid ourselves of viruses would be to join them—viruses—in the land of the not-living. I'm not ready yet. Until then, we live with.

....

Not all viruses cause a plague, but plagues will occur—every so often—on a viral planet like ours.

How do we ethically live with a plague, and how will we ethically live beyond it, if we do? Hilton Als, in his book of essays *White Girls*, wrote about the psychological effects of surviving the HIV plague years as a gay Black man in New York City. In the essay "Tristes Tropiques," Als describes spending his adult life haunted by the black garbage bags those who died of AIDS were stuffed inside in the 1980s, in particular his first love, his beloved, K.

"The truth is, I have not been myself lately, and for a long time," he wrote. In his apartment, "you'd find the same dull isolated crud you'd see in any number of apartments occupied by men who could not move on from AIDS. In any case, moving on was a ridiculous phrase, given the enormous physical memory of your loved one being stuffed in a black garbage bag like [an] imperfect piece of couture. So much time and effort had gone into creating this dress or that person, but it was imperfect, and its imperfections could contaminate the rest of the line, bag it up fast, seal it off, move on."

He could never tell another man he loved that he loved him: "In all the years I loved him, I did not say I loved him, or, more specifically, how I loved him. If I did, wouldn't that end up in a garbage bag, too?"

The HIV years taught him that they wouldn't die—the object of your affection, your love—if you didn't touch them. How can one unlearn this profound bodily truth even if it's no longer true, if HIV is no longer synonymous with death?

What residues will COVID leave behind? For the COVID-19 plague years, our grandparents were less likely to die if we didn't visit them. A new date, a new potential beloved, was safer if we met only by FaceTime.

"But who doesn't long to be mundane?" Als asks us. Look up. What was dangerous, when we consider COVID-19? Shared air. A breath is an offering to the night's sky. The breath, and so the sky, the very thing we must worry about. On the subway, summer 2020, I looked up at a man's beautiful face and smiled at him, until I remembered I shouldn't have been able to see his face, and my body pulled itself tight, and it inched away before I even thought about it, and then his mouth was an infection, and then I stood and walked to the end of the car, and then my mind calculated how much of his air from his handsome mouth could still reach me at 20 feet away, at 40, as far away as I was, as far away as I could be. Every meetup, every pod dinner, could bring death, and it was still worth it because a year without friendly touch wasn't a year I could survive. Yes. Yes, a needle did pierce my arm and then, over weeks, my body lost its reflex of pulling away from every mouth it could see. I would feel, one day, I did feel, the pleasure of lips on my lips in a bar, drunk, and it felt fine, thanks to a vaccine.

See what a vaccine can do. May 2021. For the first time in 15 months, I am in a line outside a bar. I'm wearing a sheer, extra-long shirt and black jeans and a black leather jockstrap underneath. I won't take my pants off, no. I will loosen them at the waist and have the top of my jock and the top of my ass just visible. It's my first time back; we're moving a bit slow. The line isn't long, and it moves quick enough. When it's my turn at the front, I hand my ID to the first man at the door and show my phone to the second.

My vaccine app, my New York State Excelsior Pass, proves that my lungs and mouth and all the rest of me are safe enough to enter. Four months ago, already, I moved into a special class of New Yorkers: the fully vaccinated.

I'm more squeamish than I was the first time I stepped into the Eagle, New York's gay leather bar, known for blow jobs in its dark corners and how the sign above its urinal trough "No Laying In Trough" isn't 100 percent effective. Then, I was scared of the sex and sexuality of this new and unknown place. Now, I'm not nervous about being COVID-19 safe in 2021; I know just how well my two doses of vaccine protect me. COVID-19 cases in New York are low and still dropping. This, here, this moment is as "safe" as any can get, and I feel whatever the risk, it is worth the pleasure I will get from being here, among my homosexuals, once more. But I still don't feel right being out when everyone in our city, nation, and world don't yet have access to my rare safety. After all this time, after all this grief, I also know that I am allowed to feel the small pleasure of kissing a friend of a friend on the cheek, good to see you again, after all this time, how have you been, about as good as you can expect. Yes indeed, no cheap pleasures in a pandemic.

Andrei has his arm around my shoulder and Devon has his hand in mine, my little family.

The body keeps the score, whether or not we want to remember. Look up. On the roof of The Eagle, you can look up and see a small patch of dark sky, no stars, and buildings all around reaching up, one topped with a water tower, like so many buildings in New York City, which I am only just now realizing looks like a bacteriophage binding to and infecting its host bacterial cell, the building. If we remember, it will be easier on our bodies, our minds, and each other. Only months later, with surging cases of the delta variant in my city, the fear of COVID-19 in a crowded bar returned and—most nights—the fear or the risk became more pressing than the pleasure to be found there.

None of this is leaving us any time soon, even if the virus can come and go. None of this is leaving us any time soon. I will always remember how his arm felt around me as I pulled down my mask in a bar half full of faggots to take a long sip from a beer and, for the first time in so long, too long, smile at a boy, before covering up my face again to hide my mouth from the dangerous air.

....

And SARS-CoV-2 and HIV will probably outlast my own little life on this planet. In the history of humanity, only one virus that widely infected humans has ever been wiped off the face of the earth. Almost every other virus that once infected humans infects us still, even if those infections are few (like polio, like rubella) and are generally rare and isolated and controlled. And it's possible to almost get rid of a virus, like measles, just to get complacent with the tools used to keep its spread rare. If we stop actively pushing against viral spread, a virus can reemerge up until the moment that it exists nowhere, in no one, in no animal, on our planet.

A highly effective vaccine against smallpox was needed to eradicate the virus from the planet, but the vaccine itself was only the first step. Versions of a smallpox vaccine (including using cowpox as a more poorly infectious but sufficiently related virus) existed in the eighteenth century. The technology of smallpox variolation—the precursor to our modern vaccines—was carried to the Americas by an enslaved person—Onesimus—owned by the man given the credit: Cotton Mather. Onesimus told Mather of purposefully rubbing infectious pus on his skin to protect against subsequent infection, which Mather then used to save lives during a smallpox outbreak in Boston.

Smallpox is a viral infection that causes skin lesions and can be deadly, especially for children. It can also leave the skin permanently damaged.

Smallpox eradication began, as a global project, in the 1950s. The vaccine against smallpox was a live attenuated virus vaccine, meaning that a poorly infectious version of the smallpox virus itself was used to vaccinate, and this virus wasn't killed in any way in the lab prior to its use. Live vaccines are often more effective than killed virus vaccines because the small amount of viral replication that the attenuated virus manages to get through activates a more robust immune response. There are more antibodies against the virus and better protection.

Eradication marched its way through the United States and through Europe, where institutions could manufacture plenty of vaccine, and that vaccine could make its way to patients with a refrigerated supply chain.

Eradication lagged elsewhere in the world. If infectious disease eradication lags somewhere, it hasn't worked anywhere. Some countries have little or no healthcare infrastructure as a legacy of their colonization and ongoing neoliberal policies, and so doctors' visits and vaccination are not a routine part of life.

The WHO had to work with community organizations in countries where the trust for Westerners arriving with magical medications was nonexistent, and for good reason. Scientists had to develop a way to freeze-dry the live smallpox vaccine so it could travel to places far from electricity and fridges and still be active. The push to eradicate smallpox from all countries on earth, and not just the countries where health infrastructure made it a simpler task, began in the late 1960s.

Smallpox remains the only human infectious disease ever eradicated by vaccination. Vaccination is the only tool that has ever led to the eradication of a common infectious disease.

Smallpox was officially eradicated, according to the WHO, in 1980.

The smallpox virus exists today only in labs. It's now considered high risk for bioterrorism. We no longer vaccinate against the virus. I've never had the smallpox vaccine. The risk associated with vaccination, and no vaccination can ever be fully devoid of risk, are greater—now—than the risk of catching a virus that doesn't exist.

"Americans can't deal with death unless they own it," wrote artist and writer David Wojnarowicz as he watched his friends die of HIV. "If they own it, they will celebrate it, like in the air force base museum of the atomic bomb, where whole families of camera-toting tourists gather after the required ID security checks."

Most viruses that exist on earth have nothing to do with war. They're little genetic machines with one goal in mind: to copy themselves. We're larger genetic machines with a more complex system of desire, and one sometimes at odds with the viruses we live with: most of us want, most of the time, to survive, and viruses almost never get in the way of that survival, except when they do.

....

We don't always live in plague times, but we always are outnumbered by viruses on this planet by a factor that is so large that it borders, in my mind, on infinity. We always live on a mostly viral planet.

Modernity—the human possibility that came alongside and after the industrial revolution—has granted us technology that once would have seemed impossible. Modernity, too, has made the world seem safer, less dangerous, at least in terms of medicine and disease. As Michel Foucault demonstrated, we micromanage our health through biomedicine and other forms of control (for those who can access them): we see the doctor once a year, get that prostate exam, go to the gym.

But this apparent safety is a facade; there is no safe living in a human body, and we're all destined to fall toward illness and death someday.

Modernity's apparent comfort (iPhones! GPS! Electric cars! Indoor plumbing!) comes at a human cost. Where, for example, does our iPhone battery come from? I can take HIV pills that cost $20,000 a year, but I know that many across the world cannot.

It feels like no decision we can make is an ethical one. Theodor Adorno wrote, in one translation of his work, that "wrong life cannot be lived rightly." He asked "how to live one's own life well, such that we might say that we are living a good life within a world in which the good life is structurally or systematically foreclosed for so many." He wrote from a world before COVID-19. But this is not our first pandemic, our first crisis, our first plague, our first shock.

COVID-19 might be a reckoning for many, but for others, the fractures of late capitalism and neoliberalism have been and remain a preexisting condition. Those of us who are white and upper middle class (or rich) and who live in America and who have college degrees and good enough jobs to own a home can consume "ethically." We're not meant to consider who does and who does not have access to this particular ethics.

In the cruel world of late capitalism, no individual decision can ever be ethical. Nor can an ethical life be lived. Cruelty is so deeply embedded in the system (and often so invisibilized) that it cannot be avoided. It doesn't matter that you get your food from your local co-op or farm share. It doesn't matter that your cotton is locally sourced, your coffee fair trade, that you own a Prius, that you bike to work.

Local food will never meet the needs of all the people on earth; workers at fair trade coffee growers make the same as workers at nearby farms; and your Prius's engine and battery require environmentally destructive mining practices. I bike to work.

And this, too, before all our iPhones required cobalt extraction from mines that have children and adults, alike, digging in the soil with their fucking hands. "Miners are risking their lives," an *LA Times* lede reads, "to provide you with a device that lets you read this while you're waiting in line for the bathroom."

Liberal progressivism is mostly, the argument goes, about performance, guilt, and self-righteousness. It identifies problems (climate change, environmental destruction, working conditions, income inequality, pandemic preparedness), but the solutions refuse to alter the capitalist status quo and so are not solutions at all.

The solutions are theater, and the paying audience is the same as Broadway's. $800 to see Hamilton? $75 for an organic cotton tee? How much is *your* weekly farm share? We got nothing but ramps. What are you cooking *this* week?

Adorno wasn't wrong, but he wasn't right either. I don't think. It's too easy to pass from "wrong life cannot be lived rightly" to hipster passivity to pessimism to nihilism. If it's impossible to live and avoid cruelty, I might as well live well myself. Let me quit this science and writing and activism work and go work in venture capital, buy a BMW—no, a Maserati—move to the Upper East Side, buy a house (with a pool) in the Pines.

How much does trying matter? Really trying? What is a loving ethics on a viral planet in the time not just of late capitalism but of pandemic?

Queer people provide a model for that, for living rightly in a wrong world and in time of plague, and I want to believe it's possible, because it may be believing that makes it so.

The response by the Trump administration to COVID-19 was to get things back to "normal" and as fast as possible. We weren't meant to question what normal was like and whom it served. And normal now includes life alongside COVID-19. The HIV and COVID-19 pandemics may shift into new phases, but these viruses, and our experience of them, is here to stay. There is no going back.

HIV taught us better, but we're failing to listen. We've been living with HIV for four decades that we knew of. In 1984, Health and Human Services secretary Margaret Heckler promised that an HIV vaccine would be ready for testing within two years, only one year after the virus was identified. Howard Markel wrote, of Heckler's 1984 prognostication: "Several scientists seated in the packed auditorium "blanched visibly." Their reaction was understandable. After all, it had taken 105 years to develop a vaccine for typhoid after the discovery of its microbiological cause; the *Haemophilus influenzae* vaccine took 92 years; pertussis vaccine, 89 years; polio vaccine, 47 years; measles vaccine, 42 years; and hepatitis B vaccine, 16 years.

By 1990, 100,000 Americans were dead of the virus. There is no HIV vaccine, even now. But in 2021, a glimmer of hope: an mRNA vaccine against HIV proved incredibly effective at mounting a strong antibody response. The first—the only—mRNA vaccine ever on the market? The Pfizer and Moderna vaccines against SARS-CoV-2.

I imagine another shot, one not yet possible, that makes HIV drugs for HIV prevention no longer necessary as long as I live.

In 2017, for the first time in my life, and with HIV drugs in my blood, I went to bed with a boy who wasn't my only boy and I didn't use a condom and I felt fine. He first put his fingers inside, and I pushed back on him. Summer day, afternoon, too small apartment, too small bed. His long limbs fell onto the floor. I pushed back to where he could still reach my body and the part of it that suddenly become the most important, my ass. I didn't know his name. I wouldn't say that I didn't worry, but I would say that I knew my worries weren't based on risk. He pushed into me, pushing all thought out of me, even what I knew of a virus. I breathed in, I breathed out, I felt good, and we started using one another to feel even better. There was so little risk as to be none. He came in me, that boy, on some lazy Sunday afternoon, and I felt fine. Fuck fine. I felt great. Over the next two years, the worry even receded from my mind. I was having sex *without* HIV for the first time in my life precisely *because* I had HIV drugs in my body.

Decades into living with HIV, the virus had only just begun to recede from my conscious mind. Those decades included so many living and dead activists and scientists. Drugs were found and others were made. Vaccines were tested and failed. Treatments were developed and then made too expensive and then made less expensive and then—finally—available first in America and in Europe and then finally in South Africa. They're still not available enough.

"Death comes in small doses," wrote David Wojnarowicz about the HIV living in his body. "And my anger is more about this culture's refusal to deal with mortality." We want people to live long and die quietly and out of sight. "My rage is really about the fact that WHEN I WAS TOLD I'D CONTRACTED THIS VIRUS IT DIDN'T TAKE ME LONG TO REALIZE I'D CONTRACTED A DISEASED SOCIETY AS WELL."

We live in a diseased society and have for some time. For forever. Our history in flash-forward is one of Indigenous genocide while Black slavery then Civil War then failed reconstruction then Jim Crow while Triangle Shirtwaist Factory and its fire then the New Jim Crow while Stonewall and its riots then DADT and always imperialism always invasion Bush said "MISSION ACCOMPLISHED" always rural poverty always urban poverty always low minimum wage no health insurance no unemployment but what about welfare queens and have I ever told you about the KKK burning a cross in my hometown—rural Washington—in 2003, yes 2003, yes the liberal Pacific Northwest, yes, just after I moved away to college. A Black family moved into the wealthy neighborhood built around a gaudy golf course and a few high school students burned a cross on their lawn to show them they weren't welcome.

Sick.

Well: The history of resistance too, of small acts of defiance alongside the ones we wrote down, like Indigenous wars for their own land and reconstruction and voting rights and Black power and Yellow Peril and Stonewall and Queers for Economic Justice and the ones who wrote *themselves* down like Malcolm and Martin and Fanon and Assata and Angela and Toni Cade Bambara and Hilton and David and Susan and Paul Monette, too. Well is having raw sex with a stranger, a man if you want, and just enjoying it, no fear. Well is the model that queerness offers, including mutual care that doesn't require kinship, family building outside of how capitalism defines family, pleasure as something joyous to be made and shared.

Our society, also, a quantum state of sick/well. It may never work, but I want to live a life dedicated to pushing the needle in the direction of the well, not for my body—a true lost cause—but for us, our world, and the people we'll leave behind when we go.

....

How do we try to be well, as well as a society of fragile bodies can be? Was Adorno right? Is living ethically a lost cause? I want to look back—again—to look forward at our individual ethics, our lives, and how we live. But first, politics in the collective sense. I'm a scientist, researcher, and advocate after all. I've spent months on the scientific response to COVID-19, and science doesn't get done without a huge political investment, both financially and in political will.

Science advocacy from the HIV years gives us a path. We need basic research into the virus itself. We need drug development to move aggressively on all promising drug targets. We need to figure out how to get people to trust the vaccine. We need to invest in peer- and community-led vaccination and health programs. We need to make it free for the billions of humans on our planet because everyone has a right to life, to breath. We need to start seeing housing, jobs, and income as a part of our health policy—it's hard to get a vaccine to people without a home address.

We now have COVID-19 vaccines, Amen! How long will it take until every human on this earth has access, and how many lives will we lose as they wait?

Universal healthcare would help. Universal basic income would help. We need a social safety net that will protect our vulnerable and also make sure they can protect themselves. If people need money so badly that they work when they're sick, COVID-19—or the next viral disease—will spread precisely because we're not taking care of everyone in our population. We forget too often that our entire population, and the population of the world, is a single community.

Viruses ignore borders. Viruses know that borders are made-up lines, they exist only in the human imagination, they're vestiges of a colonial world, deadly and violent, yes, but also meaningless and arbitrary. Vaccines stop at borders; countries vie for them, markets trade money for them, people wait for them, people die of COVID-19. If SARS-CoV-2 is stopped everywhere but a few under-resourced countries, it will come back here. It's in our nation's material interest now to care for the body of every human body, since we are all connected by the spread of a virus.

Ensuring worldwide vaccination is a moral imperative because vaccines save lives, and lives are precious things. Caring for the body of every human body has always been of moral interest, and we ought not need a virus to remind us of that.

....

When was the last time a plague ended? Well, in 1996, Andrew Sullivan wrote a lengthy essay—"When Plagues End"—about his own experience of living through the 1980s and 1990s as a gay white man. At the beginning of the essay, he wrote about caring for friends as they left the earth, holding hands on a hospital bed.

But for Sullivan, with the successful treatment of HIV via multiple-drug therapy, AIDS was over. The end of the plague. Life was finally *normal* again. The memory was like the KS lesions he saw on a man who was refusing HIV treatment (he assumed) at the Black Party. Another reveler asked that man to put on his shirt: KS lesions were bad memories, not welcome at the party.

For Sullivan, the HIV plague years might well be hidden, too, for the sake of "progress."

To be fair, at the beginning of his essay, Sullivan acknowledges that "the vast majority of HIV-positive people in the world, and a significant minority in America, will not have access to the expensive and effective new drug treatments now available. And many Americans—especially blacks and Latinos—will still die."

And yet, later in that same paragraph, Sullivan states that HIV "no longer signifies death. It merely signifies illness."

But what does it mean to be ill if illness doesn't imply a closeness to death? And it no longer signifies death for whom? The rest of Sullivan's essay answers that question: upper-middle-class white gay men in New York, San Francisco.

Most disturbingly, Sullivan argues that the HIV crisis finally made straight America view the gay man as A Man, worthy of respect.

Perhaps now we can all be equal, he hopes: "Plagues and wars do this to people." Go on . . .

"They force them to ask more fundamental questions of who they are and what they want. Out of the First World War came women's equality. Out of the second came the welfare state. Out of the Holocaust came the state of Israel."

Andy, women's suffrage was for white women, the welfare state post-WWII was for white people (in the South this was enforced by Jim Crow, in the North by federal law), and the state of Israel was itself borne of genocidal displacement and ongoing colonial occupation of Palestine. But, go on . . .

"Hovering behind the politics of homosexuality in the midst of AIDS and after AIDS is the question of what will actually be *purchased from* the horror. What exactly, after all, did a third of a million Americans die for? If not their fundamental equality, then what?"

This is the limit of capitalist and consumerist thought. The suffering and loss of HIV, for Sullivan, gives the gay community *capital* with which to *purchase* "equality."

James Baldwin explains this mentality perfectly in an interview published in 1984: "I think white gay people feel cheated because they were born, in principle, into a society in which they were supposed to be safe. The anomaly of their sexuality puts them in danger, unexpectedly." HIV had just been "discovered," but in and before the 1980s, there were many preexisting ways that gay people were put to death. Baldwin continues, "Their reaction seems to me in direct proportion to the sense of feeling cheated of the advantages which accrue to white people in a white society."

And, Andy, viruses are accidents, genetic recipes for themselves. That HIV killed so many gay men was indeed a horror, but not a moral one, not one based on the way gay men lived, even. Ascribing meaning to horror can help us conceptualize the harm done, of course. But there are costs. The horror, the plague, was nothing more than a horrible accident, and gay men in New York City and San Francisco were just one of the many global communities who suffered.

But Sullivan argued that the trauma of AIDS should buy gay men "equality." So what does equality mean to you, Andy? What does gay freedom mean for Andrew Sullivan, gay man?

"Before AIDS, gay life—rightly or wrongly—was identified with a freedom from responsibility, rather than its opposite. Gay liberation was most commonly understood as liberation from the constraints of the traditional norms, almost a dispensation that permitted homosexuals the absence of responsibility in return for an acquiescence in second-class citizenship. This was the Faustian bargain of the pre-AIDS closet: straights gave homosexuals a certain amount of freedom; in return, homosexuals gave away their self-respect."

For what it's worth, the pre-AIDS closet included a lot of physical violence against queer people, which is not what I consider freedom. For what it's worth, not buying into the capitalist project of a nuclear family—2.3 kids, a dog, and a picket fence on Long Island—might make us both free *and* self-respecting. I feel free *and* self-respecting when my partner and I take turns fucking a top, Andy.

Sullivan describes the end of AIDS as consisting of a "survivor's responsibility": "It is the view of the world that comes from having confronted and defeated the most terrifying prospect imaginable and having survived. It is a view of the world that has encompassed the darkest possibilities for the homosexual—and heterosexual—

existence and now envisions the opposite: the chance that such categories could be set aside that the humanity of each could inform the humanity of the other."

But Andy, the humanity of heterosexuals was never much in question, was it? But Andy, maybe I found my humanity *exactly at the moment* wherein I decided to identify as a homosexual *precisely because* it placed me in opposition to the heterosexual male who had tormented me for so long, laughing at me because I was too soft and too bookish and I cried too easily and I liked sewing and baking and cooking.

I don't want to be straight. I don't care whether or not I'm HIV-negative. I just don't want to die, not yet. The HIV pandemic isn't over, and it certainly wasn't over in 1996. It might have been for Andrew Sullivan, who wanted so badly to be integrated into the world of straight white men and their power.

Not all queer people at the time viewed the virus in that way. In their resistance, we find our path forward.

....

The early HIV plague years were also a model for community care in the face of a pandemic, for collective living and loving in spite of a virus.

David Wojnarowicz wrote, "I want to throw up because we're supposed to quietly and politely make house in this killing machine called America and pay taxes to support our own slow murder and I'm amazed we're not running amok in the streets, and that we can still be capable of gestures of loving after lifetimes of all this."

Some gestures of loving. Paul Monette to his love, Rog, as he died, for example:

> Till this one night when I was reduced to
> *I love you little friend here I am my*

sweetest pea over and over spending all our
endearments like stray coins at a border
but wouldn't cry then no choked it because
they all said hearing was the last to go
the ear is like a wolf's till the very end
straining to hear a whole forest and I
wanted you loping off whatever you could
still dream to the sound of me at 3 P.M.
you were stable still our favorite word

Essex Hemphill, for example, to his friend Joseph Beam, who died:

Our loss is greater
than all the space
we fill with prayers.

It is difficult
to stop marching, Joseph
impossible to stop our assault.
The tributes and testimonies
in your honor
flare up like torches.
Every night
a light blazes for you
in one of our hearts.

Marie Howe, for example, to her brother, who died:

For three days now I've been trying to think of another
word for gratitude
because my brother could have died and didn't.

because for a week we stood in the intensive care unit trying not to imagine
how it would be then, afterwards.

My youngest brother, Andy, said: This is so weird. I don't
know if I'll be
talking with John today, or buying a pair of pants for his
funeral.

And I hated him for saying it because it was true and
seemed to tilt it,
because I had been writing this elegy in my head during
the seven-hour drive there

and trying not to. Thinking meant not thinking. It meant
imagining my brother
surrounded by light—like Schrödinger's Cat that would be
dead if you looked

and might live if you didn't. And then it got better, and
then it got worse.
And it's a story now: He came back.

David Wojnarowicz watched Peter Hujar die and took pictures of his death. David Wojnarowicz drove Peter Hujar to Long Island for a quack doctor's therapy that he knew wouldn't work, and Peter Hujar berated him the whole time and tried to take the train back to the city and made everything difficult—the dying can do that sometimes. It might be the only thing they feel like they can do, and David Wojnarowicz still drove Peter to a doctor on Long Island that he knew wouldn't help anyone. That's how he loved Peter as

Peter was dying. Small gestures of loving like driving to Long Island for nothing.

Writing of his friend Peter (not Hujar) who died of AIDS, Alexander Chee wrote that Peter is now in "a heaven where everyone dresses well and mercy means love and a man you don't know will hold your hand when you die." Gestures of loving. We are capable. "A heaven where, when there's injustice, you chain yourself to a train because you know that somewhere someone feels it. Somewhere along the spirit-chain world-mind oversoul. Someone somewhere who maybe thought there wasn't a thing called strength feels how you care enough to stand in front of the passage of a train."

We owe it to Peter and Alexander and Marie and Peter and David and Essex to stand in front of the trains our world has sent our way. I am sorry that there are so many.

The living bore witness to the living and the dead and the dying. They held hands. It's the suburbanization of America that imagines life without pain or loss or death, without poor people or people of color or homelessness or sick people. It's a Faustian bargain. These things exist; it's not moral to live in a place where they aren't visible just because one can afford it. It's not moral to only go to circuit parties where no one can attend except platinum members of Barry's Bootcamp.

In the peak of the COVID-19 crisis, we couldn't even hold hands. The very gestures of care we need to resist American atomization, aloneness, isolation were the gestures that put us most at risk.

In his memoir, as he lay dying, David Wojnarowicz wrote, "When I was small and it would rain I thought it rained all over the world but now I don't think so."

I've been to the Black Party only once, and it did rain that night. Lessons from the HIV/AIDS plague years might be learned at the Black Party, but not in the way Sullivan means. KS lesions on the dance floor are the lesson. Joy in the face of KS lesions, joy not policed by invisibility, is the lesson. Remembering and dancing anyway. Not hiding our sorrow even while we feel pure joy. Joy is not in opposition to political will and struggle. Joy is a component thereof. Let us organize our politics in opposition to this expectation, the idea that joy means to escape the world and not to change it.

José Muñoz reminds us, in *Disidentifications*, that a politics of opposition—even if it's in opposition to capitalism and White Supremacy and heteronormativity—still centers those oppressive forces as the status quo.

David Wojnarowicz wrote, "Because I am born into a created system of corruption does not mean I have to turn the other way when the fake moral screens are unfurled. I am just as capable of creating my own moral contexts."

We cannot presently live outside of capitalism. Adorno was right to remind us of the cruelty our lives now—by definition—engender. But Wojnarowicz reminds us that we *can* create our own moral contexts outside of the killing machine of capitalism. Small acts of care defy the cruelty the world programs us for. I remember cooking dinner for a neighbor. I remember biking all the way to Manhattan just to wave up at the window of a lonely friend and speak to them on the phone, but visible, too. I remember taking the call—at midnight—of a friend quarantined with fever who hadn't left their home in two weeks. I remember helping to organize a protest for COVID-19 public health. I remember showing up at a protest for Black Trans Lives. I remember laughing and crying with those who've lost a beloved. I remember dancing together in separate apartments to Diana Ross and laughing, head back. My love: I could almost feel your hand on my back as we twirled together on screens, apart in space, safe, to the music—

Someday we'll be together
Say it, say it, say it again

Someday we'll be together
You're far away from me my love, and just as sure my,
My baby as there are stars above,
I wanna say, I wanna say, I wanna say some day we'll be
together

—a drink in my hand, a drink in your hand, like before, like always.

"To turn our private grief for the loss of friends, family, lovers and strangers into something public would serve as another powerful dismantling tool," wrote David Wojnarowicz before he died. "My silences had not protected me. Your silence will not protect you," wrote Audre Lorde as breast cancer grew and then receded in her body.

Don't pretend we shouldn't speak. Don't act like we can forget. Don't pretend we musn't act.

SARS-CoV-2, even now, as I write, is passing from one human body to another, a dangerous enough transaction. Most scientists now believe it will become endemic, passing from animal to animal for good, for forever, and some of those animals will always be humans. Even now, it seems to be finding animal reservoirs, deer and rats and mice, just what viruses do when they fail to get wiped off the earth even when we kick them out of humans, like rabies. A few thousand cases of rabies a year, still, in *Homo sapiens*, because rabies will always be out there in animals, and animals will always be close to us.

The world is never the same after a plague. Maybe because there is so rarely a true *after*. There is a trauma to deal with, the black garbage bags that queers just a few years older than me were shoved inside, even as we still live with the fact of HIV still on our planet. Our COVID-19 trauma to deal with: a touch, simple as a handshake, now deadly. A kiss may seem suicidal. Drone images of mass graves in Iran like New York. The fresh, humid night air on our lips as we walk in the city, the sweaty mess of a dance floor packed with bodies, can I ever feel these things again in the same way?

Don't pretend we won't remember. The TV mount that finally broke me. The taste of a tomato I didn't grow, popping with salt, and whose life it might have cost. Driving Devon to Long Island, fresh snow on the ground but nothing but blue in the sky, to say goodbye to his grandmother. Walking into a classroom, happy to see colleagues, afraid we might kill one another. The first subway ride. The first risky hug. The pinch of a blessed needle meeting my epidermis. Andrei on the phone saying he put his mother in the hospital and now she won't come home.

I will remember. Forgetting in order to move on will come at a cost that I will pay later, I know this much about being a human. I will remember, and I invite you to remember with me.

And so we must remember this, too, how we cared for each other. How we stood up to the forces of mass death in our government and communities. In order to remember it, we have to first do it and keep doing it, even if the daily nature of our American pandemic changes. Being better. Being well, or as much as possible.

"Hope [is] a chain of submission," David Wojnarowicz wrote. Hope matters only if it prompts action. There is no hope in the status quo, we've tried and tried and tried.

"My thoughts consist of wondering if the earth will spin a little faster when my thoughts become actions." The earth spun a little faster when you lived, David, and faster still when you died. I hope you see some of your thoughts in my actions, too, and that the world spins faster for them.

It's sunny here where I sit. Is it raining where you are? The TV stand that almost broke me. Remember that the rain won't last forever and that it will never cover the entire earth. A fresh tomato on the tongue. How I held Devon in my sleep. No matter how low, how thick, how dense and gray-black and everlasting the clouds might seem, the sun is on the other side of them, a memory now, but that doesn't mean it's not real.

11

On Evolution

Change Is God

"Who would be bothered doing science if it weren't erotic?"
—ANNE CARSON

Once, waking up in the arms of my then partner in our West Village rent-controlled apartment, our dog Winston stretched out between us. Once, in college, playing clarinet in the symphony, lost, entirely lost, in Shostakovich. Once, writing a good sentence, one I knew—even as I wrote it—encapsulated a complicated and hard and true thing that I'd been thinking about for months. Once, on the dance floor. Once, reading the D. A. Powell lines "now the mirrored rooms seem comic. shattered light: I once entered the world through dryice fog / not quite fabulous. just young dumb and full. come let me show you a sweep of constellations: 16, I was anybody's." Once, young dumb and full of cum, kissing a college girl with my college lips outside in the Minnesota wilderness in Minnesota spring. Once, on the dance floor, sober this time. Once, in Paris. Once, in Spain. In Paris, a bagel. In Spain, a bathhouse. Once, in my humble hometown, back home visiting, having wine outside in full view of the setting sun. Back home, a different time, eating lunch (a grocery deli sandwich with that mayo that tastes like only those sandwiches have, where do they get that mayo?) on the summit of a long, steep hike, looking out at the bluegreen ocean of mountains. Once, running a 10K. Once, hot and dry and thirsty, at my first sip of cold water, and once, again, with cold beer, doing as my study abroad host father once taught me: pouring it quickly into a cold, tilted glass, slicing off the extra foamy head with the back of a knife, and taking the first long sip.

Many moments in my life I have wanted to freeze, to extend out into forever, a continuous line of something like pleasure, or maybe it's a fullness of my senses, or whatever transcendence might mean. In those moments, for a brief instant, I felt a desire to feel that way forever. If I was conscious of anything, it's that I wanted nothing to change.

Yesterday, tasting the first bite of a tomato I'd just pulled off a plant I grew myself, lightly dusted with kosher salt.

I didn't succeed in pausing time as the tomato spilled its insides—acidic, sweet, salty, bright, with a flavor, too, of earth, of the dirt—on my tongue. The molecules in my food opened taste receptors that allowed my brain to feel, and my brain did feel. But those same receptors shut, and my brain returned to baseline, to this same old world where every day brings a million indignities, most of them too small to complain about. We try anyway. I know that the proteins that open in my brain to let me feel pleasure cannot stay open forever; that would bring death, or insanity. So back to this world I return.

In the 1993 novel *The Parable of the Sower*, Octavia Butler wrote about a teenage girl living with her family in California in the 2020s. The country had been rocked by global warming; water cost an arm and a leg (and one risks those things anytime one goes to fill up water bottles at a roadside station); neighborhoods wall themselves off and build citizen-militias for protection; the protection is never enough. The narrator, Lauren, builds her own religion based on the life around her, with the central tenet "God Is Change." Life is change, by definition, she argues, and pain comes when that change is resisted:

Why is the universe?
To shape God.

Why is God?
To shape the universe.

All that you touch,
You Change.

All that you Change,
Changes you.

The only lasting truth
Is Change.

God
Is Change.

I wonder if, in our 2020s, Lauren would have been a molecular biologist.

....

I am not myself today, and for a long time.

I evolved. Evolution: the change in allele frequencies over time. An allele is just one version or another of a gene, and a gene is a recipe written in DNA. Evolution, then, is just genetic change over time.

Without evolution there is no life. In humans, evolution occurs not within an individual, but within a population. Change, evolution, takes generations. Without evolution there is no future.

I am more than my DNA, certainly, but I start there. My cells all derived from one cell that was fertilized in my mom, half her and half my father. The last time I was the same as myself was in that one cell, 39 years ago now. From that cell, my brain—some three weeks later—birthed itself, my cells crawling to make the extended structures that, when electrified, make me think and feel and taste—on an August day—a tomato I grew myself.

I know a professor of developmental biology who shouts whenever he talks about the aphorism about which came first, the chicken or the egg. "The egg!" he shouts. "We know for sure. It's the egg!" What he means is that in sexually reproducing organisms like humans and chickens, evolution happens between us and our gametes, our sperm and our eggs. Once the egg is fertilized, we're more or less stuck with the DNA we're born with. "It's the egg!" I hear him shouting even now.

The DNA in the first cell—half from each of my parents—was my start. That DNA is now in every one of my cells, or nearly so. DNA isn't copied—replicated—by magic or mayhem. We call the machine in my cells that copy my DNA the replisome. The replisome is a careful machine, and it's nearly perfect. The *nearly* is the rub, the danger but also the fun of life, the reason for life itself. Each time my DNA is replicated, a mistake or two is made.

At the levels of my molecules, even, I'm not myself. All my cells are different by a letter or two. Mistakes made by the replisome are known as mutations: differences in DNA sequence from one cell—the parent cell—to its offspring. These differences, these mutations, are the beginning of evolution, and they make evolution inevitable.

What I'm saying, my dear, is that I've changed over time. This time I mean it, I promise I have. I'm not the human I used to be, that single-celled thing who couldn't see or speak or feel. I'm new and newly made. From the one cell I was once, I've gained so many superpowers; I can speak and taste, I can run and sing. All this required replication. No replication without change, all the way down at the most fundamental molecule in life, change written right there in my own DNA.

So what do all these mutations do, these errors that aren't errors because the only way to stop them from coming is to die?

Mostly, nothing. Most mutations won't even be in or near genes. Their location and molecular identity are as they imply: random. Our DNA is big, and mostly a single change doesn't matter much. One different letter out of 3.2 billion? Not much of a difference at all.

The exception is where things get interesting. Most mutations that do something are bad, as mistakes tend to be. They might disrupt the function or expression of a needed thing, a necessary protein, like a protein required to give a cell its shape, a protein required to turn sugar to energy that the cell can use, a protein required to make other proteins, any protein the cell can't do without. The cells carrying this type of mutation will die, and so the DNA sequence will die with them.

They'll be selected against and disappear. Mutation is evolution: the creation of new alleles, new versions of a gene. Selection against a new mutation, too, is evolution: change in gene frequency over time, as this new version will not survive.

Very, very rarely, a mutation might do something new and interesting. It might make eyes brown instead of blue. It might make us 5 foot 5 inches instead of 5 foot 7 inches.

These mutations might survive, might persist, might become more common. We can change, we humans, but we do take time. Our generations are decades. As you know, viruses replicate too. And they can make many thousands of themselves in as little as a few hours.

Our rate of evolution depends on how many mutations we make per generation and how long a generation takes to be made. Our mutation rate is due only to the machinery that makes us. Humans are so careful, we mutate rarely, one mistake in a billion. HIV and flu have replication machineries that are sloppy. What's more, they have no way to fix the mistakes they make. Almost every HIV virus that's produced, and every flu virus too, even though they're small, only around 10,000 letters long, will have a genetic difference, a letter apart from the norm.

The majority of these mutations that have an effect will kill the baby virus, but who cares? Who cares when you make thousands of offspring in a single generation, like these viruses do. They have 100,000 offspring, so what if 1,000 are stillborn. What matters is that most are not. And 100,000 offspring can be made, by a virus, so quickly.

Their evolution, viruses, is at warp speed to the living.

What matters, too, is that viruses that live on are diverse. They're different from one another. Evolution is not an absolute; conditions change, even for a virus. And when conditions change, a version of a gene that might have gone away a generation ago may become essential now. This is the genius of evolution: it's not searching for an answer, it's allowing for constant, endless, needed change. Evolution never arrives at an answer. If it did, and the conditions changed, it would be stuck, and stuck things die. Evolution is simply a capacity to ever change. How queer.

It's 2022. We sequenced most of the 3.2 billion letters in the human genome 20 years ago. Today, we can sequence the 30,000 letters of RNA in the SARS-CoV-2 genome many, many times over in a single day. Those sequences will tell us how many mutations there are between two viruses, the one infecting me, for example, and the one infecting you. The one in China and the ones in Taiwan and Italy and Germany and New York and Seattle.

Evolution is based on random mutations and their effects; randomness, even now, is in charge of our future. Change is inevitable. Change is God. The virus can change in its very genes. Change is life. We have to change with the genetic lots we've been dealt, meaning we have to change not our DNA but our behaviors. When we build new technologies—like vaccines—that change who the virus can infect, we shift the possibilities of viral evolution, too. If vaccines prevent infection—and they seem to—they will then prevent new mutations and therefore slow evolution. No new variants, then. So, what that change means depends on us—our action—and nature, and so on random chance, which is how mutations arise, and selection, which is how they grow or shrink. We write this story together.

....

It was spring 2017 and I was feeling fine! I was walking outside on one of the first warm days of the year. It must have been April. My headphones were in. I was smiling.

Go to Google and type in "LCD I can change." YouTube. Connect to a speaker if you've got one. Go ahead, I'll wait. Dance. *Open your arms. Dance with me until I feel all right.* Dance with somebody if you're lucky enough to have somebody.

In my almost-youth, one of my boyfriends dumped me because—he said—my anxiety (which he called my mental illness) made his life too hard. Never mind that my peak in anxiety was caused by two very real world events: the election of Donald Trump and this boyfriend's decision to move out of our apartment and across an ocean to take a job in London for 18 months.

I imagined then that if I got my shit together, anxiety-wise, he'd have no choice but to take me back. I went to therapy again, after taking a couple of years off. I started taking Lexapro and quickly added on Klonopin as needed for the anxiety attacks that grew out of sleeping on my own in a bed that used to belong to us.

It was spring and it was 2017 and Donald Trump was president. I'd been single for three weeks. It wasn't lost, not yet. He could take me back. I'd been on Lexapro for a week, maybe, and I knew the drug made things worse before it made things better, but I was promised by my doctor and therapist and mother and sister and three friends who all took the drug that it would eventually help.

We can change, as people. We can stop drinking and start working out. We can learn new languages and cook for ourselves. We can be kinder to our friends, nicer to our mothers, even if it's hard work. We can go to therapy.

I was trying to change, to be less afraid of futures I couldn't control. I was trying to change in time to keep him.

I left work—a research lab then, at NYU, near Washington Square Park—and it was a sunny day and I was walking down 4th Street in the heart of the village. I was listening to this song on repeat. I'd been on Lexapro for one week. If I changed, he'd take me back. I felt . . . good for the first time in weeks. Had the drugs worked that quickly? Had I taken a Klonopin and forgotten about it? No! Maybe I was cured! I wasn't anxious anymore. With each step I took I was more certain. Ten steps not anxious. I was a new man. Now 15. *And I can change, I can change, I can change, I can change.* Now 20. He'd have to take me back! *I can change, I can change, I can change / If it helps you fall in love (in love).*

The next week, the Lexapro would turn my skin red in splotches, make me shiver even as I sweated with heat. At times, I had to lie down and close my eyes to keep from falling over. Just lie down with eyes closed. After a month on the drug, these side effects finally waned. Were they side effects or aftereffects of my breakup? Who could tell? I had headaches and blurred vision and I had trouble focusing at work and I couldn't eat but I had to eat or I would dissolve into yet another panic attack.

That spring day, in the sun, I made it to 100 steps, no anxiety. I made it all afternoon. I was convinced that I'd changed, and that I'd done it in time. I would change, yes, but not in the way I wanted to then, and anyway, the time for a change to save that lost love was long past.

....

Me? I've always hated change, even when I longed for nothing more. Even when I was a too-headstrong towheaded high school student who needed to see more than his tiny farming town could offer. It's hard to even remember that child and to believe that I was once him.

"I think we are well advised to keep on nodding terms with the people we used to be, whether we find them attractive company or not," Joan Didion wrote in "On Keeping a Notebook." "Otherwise they turn up unannounced and surprise us, come hammering on the mind's door at 4 a.m. of a bad night and demand to know who deserted them, who betrayed them, who is going to make amends."

In my tiny hometown, freshly 18, I cried and shook, my first panic attack, when I had to mail off the deposit to go to college. I had two choices, checks made out for each because one of the checks, we knew, wouldn't be cashed. Left hand, right hand, which would I choose? Providence? Northfield? I cried for the choice, for the change that I knew was coming. What if everything went to shit? What if, after a semester or a year, I ended up back here, just like everyone else? A tractor drove down the main street of my hometown behind me, and I had to laugh through tears I tried to hide and so cried quietly. Yes, back home, boys—little and big—don't cry.

I'd put it this way: Bad memories are like broken glass. You clean it up as best you can, but—two weeks later—it's never a surprise when you, barefooted, cooking, land on a shard, a half inch long, that somehow avoided the broom and vacuum and mop and hands and eye.

Left hand, and I drove home still crying. I'd made a choice; the change was well on its way. Fifteen years later, I dated, for three dates, a boy who graduated from my other college, the right-hand college—Providence—the year I graduated from the left. A shard of glass when I found out. What would my life have been, that other me?

....

In general—at the molecular level—humans don't adapt to their environment. Sure, I might learn something, have and hold it in my memory, and act differently the second time I meet, just for example, an orange-hot stove. As ever, in biology, there is an exception to this rule. As ever, in biology, the exception to the rule is where the magic happens.

We can stop drinking and start working out.

None of these changes will be written in my DNA. When a human learns or remembers, it's thanks to changes in our brain, in the expression of genes, but not in the letters of the DNA itself.

We think about human evolution in regard to viruses like this: A virus moves into a population of humans who've never had that virus before. Some of those humans die. The ones who don't die are the ones who have kids. Those kids might inherit resistance to the virus. The virus then falls out of the population, or evolves—itself—to still infect people but probably not kill them.

Individual human beings might mutate, but we don't (often) evolve. The land where the rules of biology falter is one of great beauty and mystery, as if those two things differ, for us biologists.

The first exception: I can change. Evolution does of course

occur in our living bodies. Cancer is the easiest example to consider. With cancer, randomly made mutations grant cells the ability to grow and grow. Mutations that help these cells grow bigger faster better stronger are selected for, and turn up in the eventual tumor. We can, with DNA sequencing, trace this evolution in real time just like we can watch mutations in a virus arise in a global population of infected individuals.

Cancer cells evolve inside of us, and they are made from us, and they can kill us.

Luckily we have another set of cells that can evolve too: our immune system.

Immunologists generally classify two types of immune responses: innate and adaptive. The innate immune response is always present and ready to go, and it recognizes general patterns of infection, like cell death (viruses can kill cells, after all) and molecular oddities typical of infection (the sugars on the outside of bacteria, the double-stranded RNA that our cells never make but viruses do).

When a pathogen we haven't met yet, like SARS-CoV-2 or HIV or CMV or herpes, initiates an infection, the innate immune system is alone for about a week, the only thing there to find the virus and try to get rid of it.

But the innate immune system cannot learn. It will never *change*.

Certain cells (macrophages and dendritic cells) of the innate immune system, though, will suck up the bits of dead virus it finds early in an infection and carry those dead bits back to the lymph nodes. Here, a magic happens. The macrophages will present these bits and pieces of the virus (or a bacteria or a cancer) to cells that can learn, cells that can change.

Cells that will evolve. Cells that will remember.

An adaptive immune system. Bits of virus are kissed off and onto the surface of T cells and B cells.

T cells have specific receptors for a little piece of a virus. They will recognize this virus and (probably) only this virus. For each virus, a T cell will fit it like a lock to a key.

B cells make Y-shaped proteins—antibodies—specific to a virus or bacterium. The recognition is tight and specific. Antibodies all look like each other, Y-shaped, except at the two tips of the Y, where there is a variable region, each antibody different, all its own. This variable region will recognize one protein from one virus, like the spike protein from SARS-CoV-2 or the gp120 protein on the outside of HIV.

But there are a million pathogens out there, a million viruses and some we don't know about—or that haven't even evolved yet. And yet our body—through its T cells and B cells—is able to adapt even to these viruses that don't yet exist. How?

Through mutations, of course. In this case a particular type of mutation called recombination: a whole-scale swapping of DNA. A mix-and-match. Each of the T cell and B cell proteins that need to recognize anything possible that could make us sick is made from one each of a protein section called V and a protein section called D and a protein section called J.

We're gonna do a little math, so follow me here. Only when these three sections come together can a T cell or B cell be made. But there are options: 44 options for V, 27 options for D, and 6 options for J. They're chosen at random, pulled like lottery balls out of a spinning cage. So there is something like 30 billion possibilities, and they're more or less all made in the hopes that just one or two or three of the possibilities will be able to recognize any virus we'd meet, including this one, this new virus, SARS-CoV-2.

So, honey, we have options. We have options for the CMV virus we meet before we're 2 years old, the herpesviruses we'll meet our entire life (so kiss me, darling). We have options for a coronavirus currently in bats that hasn't yet made it into humans. We come prepared. We have evolved to have our immune system evolve as well.

Say you were infected with SARS-CoV-2 a couple of days ago. The infection in your nose and throat and lungs was met by the innate immune system, double-stranded RNA a signal that a virus is afoot. Macrophages head to the scene, where they scoop up some dead cells and some virus. They float off then and head to the lymph, where they meet naive T cells and B-cells that have yet to meet their perfect mate, their match, the key to open up the door they are.

And, hopefully, a T cell and B cell will exist that fits the new virus now in us. And what's more, the fit doesn't have to be perfect, not yet.

But why? Because the adaptive immune cells—even after recombination—can do something no other cells in our body can: mutate and evolve. They have special machinery to make mutations in this one tiny region in all their DNA, right where the variable regions are in the proteins that recognize a virus or bacteria. We still, to this day, don't understand how this occurs, how mutations can be induced so locally, just here, without mutating the rest of the DNA, which could cause cell death or cancer. Mutations are dangerous and have to be used in just the right way.

We find a loose fit between T cell and virus, and then mutate. Again, most of the mutations will do nothing. But a small few might make things better, tighter. And the T cells and B cells with those particular recognition proteins will grow and grow and grow. They're selected for. They get all the good juice, those good signals. This is the precise thinking behind the need for two or three vaccinations with Pfizer and Moderna: The first shot will activate a general immune response to the SARS-CoV-2 spike protein; the second shot (and the third, and any more we take after that) will lead to fine-tuning via mutations in the antibodies and T cells tightening their recognition for the virus. The molecular, often enough, becomes medical.

This evolution, selection, and cellular expansion within us takes days; this is why the innate immune system is on its own for almost a week. Only then can the adaptive immune system, our T and B cells, arrive, after evolving and dividing to proliferate, to help. If we're lucky, it might not have been a virus that picked out the cells and activated them against a future infection, but a vaccine.

Octavia Butler's Lauren was a molecular biologist, I know: "Intelligence is ongoing individual adaptability," she wrote. "Adaptations that an intelligent species may make in a single generation, other species make over many generations of selective breeding and selective dying." Our intelligence is molecular: it's our adaptive immune system's ability to evolve and proliferate. Our behavioral adaptations arise through memory and learning, neither of which rely on changing our DNA.

There's nothing more normal than meeting a virus; we do it every day. Now—at 39 years old—most viruses I meet will look just like something I've met before. So I have antibodies and T cells ready to greet them when they arrive. The viruses I carry in me already, my chickenpox and HSV-1 and CMV that I'll have forever? My body learned about them long ago, and my immune system has cells that are constantly checking on them, making sure they're keeping quiet. It's constant, this conversation. I may well have HHV8, but I don't have Kaposi's sarcoma because my T cells are always talking with the cells where this virus lives, keeping my skin clear of red cancer. I may well get a cold sore again this year if one of my herpesviruses, for a week or two, talks louder than my T cells. I may well get shingles, a reactivation of the varicella-zoster virus that caused my childhood chickenpox, although I pray I don't. I'm grateful for the chance; the only way to be absolutely certain my varicella virus won't reactivate is if I die, and the virus then dies with me.

We meet viruses every day. There is nothing more normal than that. When a virus breaks through, makes us cough or itch or bleed, there we have an aberration. Humans don't evolve, except when we do. We'll change (T cell, B cell, amen), and the next time we meet that virus, God willing, we'll speak to it softly, with a protein, with a cell, and have no cough or itch or blood or fever that day.

....

As I chop onions and garlic and kale, I put on Rachmaninoff's Piano Concerto No. 3. In 2020, after COVID-19 kept me inside to cook five times a week, I don't know why, but I began listening to a lot of the classical music I loved as a child. Then, it made me an oddball, a target for bullying, a fag. If those bullies could see me now.

I loved classical music—and novels by Austen and Dickens—in part because I wanted, I *needed*, to be different from *them*. I hated those boys for being working-class boys who wouldn't ever leave that town. No, I hated them for pushing me into a locker and spitting on my shirt. For trying to light my hair on fire. Because of them, I needed to leave my town. And so I made myself everything they weren't. They mocked me, but I knew I was better than they were.

I drove a shitty little car—a Mazda GLC—that I bought from a family friend for $200, about what it was worth. It had no A/C—obviously—and the doors didn't seal, so when I was driving in the rain (half the year in rural Washington) my left leg was constantly wet. The windows hand cranked. The car topped out at about 65 miles per hour before the doors started to shake in their frames, making the leaking water worse. Our speed limit then was 70. My parents viewed it as a sort of insurance policy for their male teenage driver: I literally couldn't speed. It had a turn dial radio that I replaced, with money from work, with a tape deck. With money from work, I bought a CD player with skip control (it ran the battery though, so, when driving, I plugged it in for power via the car's cigarette lighting port) and a tape insert to plug it into the car's audio, which, is called a "Car Connecting Pack" and is still available for $18.19 from Walmart online (one left, buy now).

In the summers, I drove down the main street in my tiny town (Olympic Place) from the high school or Safeway past the post office toward home, or the other way around, windows cranked down on both sides of the two-door GLC, blasting Rachmaninoff's Piano Concerto no. 3, which I'd tried and failed to learn to play, and head banging like it was punk rock, which, in a way, to me, it was. The other boys broadcast their country music in this way; let me broadcast my difference from them.

"I'm getting out of this place," I was trying in my own way to say. "I appreciate things those assholes could never understand."

That was me, head banging in my GLC to the music in my ears even now as I chop onions and garlic and kale in the Brooklyn apartment I share with a man.

Tonight, I wrote and cuddled our perfect pup. It's not 4 a.m. It's not a bad night. That Joseph can stay away a bit longer. Tonight I'm cooking dinner and I put on Shostakovich after the Rachmaninoff ended—we played his Fifth Symphony in college—as Devon asks about my day. While I cook, Devon turns on his disco ball, the one I gave him as a gift last year, and stars shoot up on our walls, turning slowly. A crescendo builds toward the end of a movement. I have Lexapro in my blood. This is the closest thing I've had to a club in months, this is a night out.

"Babe," I say, one hand on my hip, another holding a wooden spoon slowly stirring a pasta sauce. If they could see me now. How I've changed. My younger self would never let my younger self hold his body this way or wear these heels or call a man something as soft as babe.

"Babe?" he says.

"Come dance with me."

"I can't dance to this!" He looks at me from the corner of both of his eyes.

How I'm the same as I always was.

"You, my love, can dance to anything."

....

We can watch viral evolution in real time, and, with SARS-CoV-2 and influenza and HIV, we have. We've watched the mutations accumulate between places and people and even, a few times, within a person, too. Human evolution, the changes in us, we have to infer over time. We don't know what the next generation will bring and we can't—and shouldn't—experiment on people in the way we do on viruses, bacteria, yeast, even mice and worms and monkeys.

But we know people evolve over generations, and we know many things about how.

Pathogens, like viruses and bacteria, push us to change. This field of biology is called host-pathogen coevolution. I am the host: SARS-CoV-2 or HIV or CMV or HSV needs me to replicate. Any of these viruses—mild or deadly—is the pathogen, and it will do anything to copy itself. It's just molecular biology, though: the mistakes made in copying itself and selection for or against these mistakes.

Host-pathogen coevolution has shown that people are uniquely diverse, at the molecular level, with genes that relate to disease susceptibility. We expect, when differences are genetic, that people will have very different outcomes when they encounter a pathogen. Around 1 percent of people are entirely immune to either HIV infection or progression to severe disease, while the vast majority of people progress to a pretty profound and dangerous immunodeficiency within a few years of infection.

Some people who get SARS-CoV-2 never feel ill at all, while other people get so severely ill that they die. Almost every child who gets CMV will barely show a single symptom, but it can indeed be dangerous for a few. This is the cost of living on a planet covered in life-forms other than us; we would survive less well or not at all on a planet without microbes. We've evolved next to them; without them, we wouldn't survive at all. Yes, I know the emotional cost of this horrific randomness; I've lost friends who shouldn't have died to a virus, and they can't ever come back to me again.

And while SARS-CoV-2 might be evolutionarily stable, because it makes so few mutations, our immune systems place an incredible pressure on it to evolve. If this virus sticks around for some time, it might indeed shift in ways that matter for its biology, for how infectious it is, for how sick it makes us, for how effective a vaccine or treatment is. And the more virus there is on our planet, the most chances it has to evolve, to change, to evade.

After nearly a year of viral evolutionary stability, we watched, in real time, something new happen. B.1.1.7, which we now call alpha. Together, all at once, all over the world, one word: Variants. Mutations. B.1.351. More infectious? Yes. More deadly? Maybe. What exactly was happening? P.1, which we now call gamma. How exactly had this happened? B.1.427, epsilon, and B.1.617.2, delta. We still don't know for sure, but we think these novel variants arise not as the virus transmits *between* people, but *in* people where the virus can persist for weeks or months. People with immune systems weakened by cancer, for example, can shed live SARS-CoV-2 for months; in this time, in this person, in the presence of an immune system, even if it's weakened, there the virus likely can evolve.

And it seemed like the variants, once they arrived, came all at once. And it seemed, based on where they came from, that they weren't related to one another. And it seemed, based on their RNA sequences, that they all shared the same mutations. We call this convergent evolution: when different organisms in different contexts solve the same problem in the same way. Usually, common mutations imply that two things are related. I came from you, mom, or you came from me, my child. But convergent evolution changes all this. The easiest example is organisms who swim: compare whales with fish. Whales are mammals, and fish are fish, and whales are not fish. Like all mammals, whales once lived on land, emerging out of the sea with the rest of the soil-dwelling creatures on our planet. But the sea called them back, and they learned again to swim, but never lost their lungs, never regrew gills. If you looked just at one trait (can swim), you'd call a whale a fish. But if you looked just a bit closer (at their DNA) or looked at more and more and more (lungs and live birth and milk and placenta), you could trace a whale's history back to the land. Fish and whales solved, separately, the problem of how to swim.

SARS-CoV-2 variants from different places solved, separately, their desire to replicate better or faster or evade the human immune system just a bit more.

We don't yet know, but we will be watching—in real time—and hoping that if this virus does change, when new variants do arise, it's to act more like the seasonal coronaviruses that make us no sicker than a common cold. What will the variants do next? We don't know. Only time and luck and selection will tell us.

Our biology is the foundation of our lives, but it isn't all we are, all we have, all we can be. We can choose to protect ourselves and each other with our behavior, putting care above all else, until human immune systems—worldwide—can meet not a virus but a vaccine.

It's possible that someday SARS-CoV-2 will be just another virus like CMV, one that does us no harm. How we get there is in large part up to us: vaccines or infections are the only way forward, and one of those two options can kill.

Then, another day, it's possible that another new virus will begin spreading from human to human. As long as we cover the earth, we will meet new viruses, ones our immune systems don't know well enough to talk to yet.

Our biology is not all we are. We can evolve beyond it, not with our DNA, no no, that changes far too slow. *We* can change. Where we place our bodies, and with whom, is a biological choice, and a moral and political one.

In response to the HIV pandemic, some American states made sex illegal for HIV-positive people without certain types of protections, like condoms or legal disclosure of the virus. These laws didn't stop HIV, they just put people with HIV in jail. This is a deadly human adaptation to a virus. That HIV first impacted queer people, immigrants from Haiti, and injection drug users drove these horrific responses. Racism and homophobia made so many feel like HIV was not their virus, which meant it could be locked up and kept away.

Certain strains of the human papillomavirus can cause genital warts, and other strains can cause cancer. This virus is sexually transmitted, like HIV, incredibly common, and usually asymptomatic, causing neither warts nor cancer. We, after years of living alongside the virus, have developed vaccines against some strains of it; we never used HPV as an excuse to criminalize sex. We are trying to prevent the cancer the virus causes. These adaptations are, as I write, saving lives.

Human lives are worth saving, no matter the genes they might have or the underlying conditions. It hurts to even have to write this. Changing one person or mind is hard enough; changing culture can feel impossible. But when it comes to questions like whether we value money over life, whether it's OK to imprison someone because of a virus they carry, a change is necessary, no matter how difficult, and we need to keep repeating this—we must change—until the day that change arrives.

....

I can change. Taking Lexapro didn't win me that boyfriend back. But I have changed: I'm still an anxious person (I mean look at our world), but I won't let people around me blame my own anxiety for their actions any longer. He wanted to move abroad; I was baggage that made that difficult. I hope, if this happened again, I wouldn't beg him to keep me even as he left and blamed me for leaving. If I did, I'd have to look my therapist in the eye and tell him what I was doing; the shame of speaking it aloud would stop me, I hope.

I'm not the person I was then, in 2017. Old cells were replaced with new. Of my 43 trillion cells today, probably more than half weren't yet born then. I can be kinder. My DNA is copying itself, and it will copy itself tomorrow. Mistakes will be made. I can forgive my mistakes and the mistakes of others. My T cells and B cells, an almost countless number, each unique, float somewhere in my butter-colored lymph waiting to meet a virus, a bacterium, a cancer. I still take Lexapro. I can work harder, and better, at all my jobs, teaching and writing and loving on my friends. And loving on myself. When my T and B cells meet a virus, what will they do but grow and change? I am still anxious. I can take more time off, put my phone down, feel the earth beneath my flat feet. I'm still anxious. I can evolve. I can adapt. I can grow. Most of the viruses I meet in my life will do me no harm. I can change. My cells change in response to the viruses they meet. I'll never be with a man (or anyone) who will pathologize my anxiety about a broken world or a broken heart. I'd rather be alone, and I'd write that in my DNA if I could, a molecular tattoo that can't be wiped away. I don't regret this person I used to be, no matter how much pain he caused me. I take solace in the fact that in me, right now, there are more than a trillion cells, all evolved, all different, waiting to meet any virus present on our planet and every virus that might arise someday. I am large, yes, and I contain multitudes. It's because I've evolved. I'm working on being kinder, and I'm asking for that kindness, too, in return. I have cells ready to respond to a virus that doesn't yet exist in me. In me, cells waiting to respond to a virus that doesn't yet exist on earth. If you look closely enough, if you look at my molecules, I'm changing, even now.

Acknowledgments

First, to my family and chosen family, without whom I wouldn't have been able to finish this book: Colleen, Jeff, and Katy. My partner Devon Wright. My forever family Whitney Richards-Calathes, Hala Iqbal, Ngofeen Mputubwele, Laila Pedro, Łukasz Kowalik, and Jesse Afriyie.

Then, to the people who wrote this book with me, including Patrick Nathan and Steven D. Booth, who co-wrote "On War" and "On Activism and the Archives," respectively. To Devon, Ngofeen, Laila, and Andrei, whose generosity with the lives we shared in our COVID-pod extends, in my view, to a collaborative writing of the essay "On Private Writing." We wrote together with our actions and revised together once there were words on the page. Thank you. For your permission to write about the inside of all of our lives. Thank you. I would not have survived without you, let alone been able to make meaning from the horror and joy we simultaneously experienced together. Our queer friendship *is* family, and it's why I am still here.

To Mo Crist, the best editor I could have imagined for this book, and Alane Mason, who believed in it and invested in it. To everyone at W. W. Norton, especially Michelle Waters, Rachel Salzman, Gina Savoy, Beth Steidle, Susan Sanfrey, and Sarahmay Wilkinson, and to Matt Dorfman for this stunning cover. To Katie Kotchman, my friend and the agent for this book, who sold it, and who supported me for so many years of writing (and living). Mo, Alane, Katie, and I co-conceived the book, and their gravity pushed and pulled the writing like a force of nature. Mo, in particular: Thank you.

Thanks, also, to the editors who I worked with on these essays for prior publication: Ed Winstead at *Guernica*, Mark Gimein at *The Village Voice*, Adam Weinstein at *The New Republic.* These editors all worked diligently to make my writing and thinking stronger, and this book benefits immensely from their work.

To my friends, friendtors, mentors, and writing groups, my best and first readers and the people I write to: Darnell L. Moore (who asked me to write publicly for the first time in my life, all this is your fault), Kiese Laymon, Alexander Chee, Lacy M. Johnson, Chanda Prescod-Weinstein, Ngofeen and Patrick, Melissa Febos, Tommy Pico, Alexandra DiPalma, Kenya Anderson, Fran Tirado, Denne-Michelle Norris, Joshua Roebke, T. J. Tallie, Yana Calou, Ricardo Hernandez, Ariel Kates, Liz Lattie, Jennie Gruber, kd diamond, Seth Fischer, C. Russell Price, Natalie Eilbert, David Groff, Miah Jeffra, William Johnson, David Kittredge, Claire Atkinson, Kristen Arnett, Addie Tsai, Jon Silver, Paul Sepuya, Alex Marzano-Lesnevich Promiti Islam, Tanaïs, Garth Greenwell, Garrard Conley, Rob Spillman and Elissa Schappell (your support means everything). To my agent Kent Wolf. To Bryan Borland and Seth Pennington for having endless love for the writers of Sibling Rivalry Press. To our dog MAX!, whose playfulness gives me such joy.

To my therapist Dr. Eric. Therapy and Dr. Eric deserve their own acknowledgements, especially for writers, and especially writers trying to do the impossible thing we cannot not do: write a book.

To my virology/epi friends (Maimuna Majumder, Jason Kindrachuk, Jo Walker, Juliet Morrison, Ryan McNamara, Michael Bazaco, Nini, Muñoz, Cody Minks, Saskia Popescu, Sara Bazaco, Anna Muldoon, Tara Smith, Emily Ricotta, and especially Angela

Rasmussen). You all provided material and fact checking for this book, support for all I was working on, and, maybe most importantly, a source of unexpected friendship, a safe space to vent and grieve, and more laughs than I thought possible given how difficult everything was and still is. I learned so much from you all. To my PhD mentor Seth Darst, to microbiologist extraordinaire Liz Campbell, to my thesis committee member and virology teacher Charlie Rice for rigorously expanding my understanding of the viral world.

To the members of the COVID-19 Working Group, particularly those with whom I worked most closely (and who agreed to appear in this book): Wafaa El-Sadr, David Barr, James Krellenstein, Judy Auerbach, Jeremiah Johnson, Matt Rose, Kenneth Mayer, Christian Urrutia, Jessica Justman, Peter Staley, Mark Harrington, Charles Kind, Melissa Baker, and Allie Bohm. I'm sure I gained as much as I gave through our work, I just hope I did as much and as well as I could.

And finally, to anyone who read this far (or even just read a little bit): Thank you for giving me some of your time, for thinking with me, and for trying—if you can, if you will—to walk forward with care for all things living, to remember all the dead, and to reconsider those small viral things in between. We won't be rid of them, ever. One day, they'll be rid of us, but they won't notice much anyway.

An online notes section can be found on my website at www.virologybook.com/notes.

An online archive can be found on my website at www.virologybook.com/archive.